ENERGY
ENGINEERING AND MANAGEMENT
FOR BUILDING SYSTEMS

ENERGY ENGINEERING AND MANAGEMENT FOR BUILDING SYSTEMS

William J. Coad

VNR VAN NOSTRAND REINHOLD COMPANY
NEW YORK   CINCINNATI   TORONTO   LONDON   MELBOURNE

Copyright © 1982 by Van Nostrand Reinhold Company

Library of Congress Catalog Card Number: 81-134
ISBN: 0-442-25467-9

Manufactured in the United States of America

Published by Van Nostrand Reinhold Company
135 West 50th Street, New York, N.Y. 10020

Van Nostrand Reinhold Limited
1410 Birchmount Road
Scarborough, Ontario M1P 2E7, Canada

Van Nostrand Reinhold Australia Pty, Ltd.
17 Queen Street
Mitcham, Victoria 3132, Australia

Van Nostrand Reinhold Company Limited
Molly Millars Lane
Wokingham, Berkshire, England

15 14 13 12 11 10 9 8 7 6 5 4 3 2 1

Library of Congress Cataloging in Publication Data

Coad, William J.
  Energy engineering and management for building
systems.

  Includes index.
  1. Buildings—Energy conservation. 2. Buildings
—Mechanical equipment.   I. Title.
TJ163.5.B84C6   696   81-134
ISBN 0-442-25467-9   AACR2

To
DOODIE

Introduction

The difference between competence and excellence in engineering generally lies not as much with a difference in technical knowledge as with a difference in philosophy. The same could be said for most other professions based upon the physical sciences, because a relatively high level of technical skills is necessary simply to achieve competence. This observation is the cornerstone of the contents of this book.

The Western world is facing a dilemma today which will change all of our lives rather dramatically in the next two decades. This dilemma, since the mid 1970s, has manifested itself in various ways ranging from gas lines in many areas of the United States; to revolution, anarchy, and foreign occupation in the Persian Mideast: to economic stagnation; to the creation of a gigantic new department of the United States government; to prime interest rates in the United States of 18 percent; to an alarming devaluation of United States currency on foreign markets; and on and on. These are not unrelated incidents, nor are they "manufactured" by and for the benefit of special interest groups. The underlying cause of all of these related incidents is the very basic and simple economic law of supply and demand. The reason the supply and demand situation relating to one resource group—energy—has such an overwhelming effect upon all segments of our social structure is simply that energy is the commodity upon which our current growth in productivity (and thus GNP) has been supported for several decades. *Remove the foundation and the building tumbles.*

The relationship between the discussions of the above paragraphs is important to recognize if the energy situation is to be successfully addressed without a collapse of our financial institutions and economic systems. Not only will this effort require an unprecedented surge in excellence in engineering, it will also require an integration between the activities of the engineering community and other professions who will be addressing the problem, that is, economists, attorneys, business managers, legislators, legislative assistants, bureaucrats, and educators. No longer can the lawmakers ignore the input of the engineer. No longer can economists solve the problems of inflation by manipulating money. No longer can the businessman treat the precious energy resources casually. The only way that the Western world will economically survive the closing decades of the twentieth century will be by utilizing the available fossil energy resources in a more effective manner, that is, in a manner which will enable us to continue to increase productivity while actually reducing our rate of consumption of fossil fuels. There is no other answer. Recognition of this concept will set the stage for the twofold needs of the long-range solution. First, it will give us the necessary time to develop alternative energy sources—whatever they might be. And second, it will provide the knowledge of and experience in the methods for more efficient or effective use of energy which will be of paramount importance in the overall economics of utilizing the future alternative sources (they *will* be both precious and costly).

This book is a collection of various articles, periodical columns, and papers written by the author. Although most of the chapters were originally intended for an engineering readership, many are more philosophical than technical, and the technical content of others is quite elementary as presented. The technical complexity of some of the writings was intentionally limited in the hope that this would assist in the necessary interaction between the engineering community and those other professions addressing the energy problems. Thus the title of the book *Energy Engineering and Management* . . . was selected in the hope that its use and its message would not be limited to those practicing, or preparing to practice engineering.

The content of many of the topics was germinated in discussions with engineering students. Because of the ever-increasing vastness of the technical substance required in engineering curricula, the present-day student has little time to develop the ever-important philosophy of design and other elements of engineering application. As a result it is found that as the technical burden has been increasing, the ranks of competence have been swelling while the ranks of excellence seem to be diminishing. Thus, at a time when there is an unprecedented need for engineering designers and educators of superior skills, too many of our graduates are seen to lack the excellence to fill the needs. This gap in supply and demand is second only to that in the fossil energy markets (but not unrelated). The solution to this problem is not more engineers, but engineers educated not only in the technical routines but also in the philosophy of engineering. To this end a good portion of these writings have been directed. Although the specific discussions in most cases relate to building systems engineering, the subject matter is applicable to all areas of mechanical engineering—indeed, all areas of all disciplines of engineering.

Some of the chapters appear, by their title and content, not to relate to "energy" per se. But considering that the definition of mechanical engineering "is the applied science of energy conversion," any and all aspects of the study, analysis, and performance of mechanical systems relate directly to "energy." Perhaps the most detrimental concept to the control of energy waste is to try to approach the problem independently of the tried, true, and established channels. Those who would approach a curriculum of "energy engineering" independently of mechanical engineering, or such subsciences as solar engineering or nuclear engineering as independent disciplines will most assuredly fail. The most effective area in which to channel all of our energy-oriented activities is in the basic area of mechanical engineering. Conversely, if we are to be successful in these efforts, *every* segment in the study of mechanical engineering must be considered as a subdiscipline of energy conversion. Thus, any topic or discussion relating to mechanical systems concepts or designs, is in fact, a discussion of energy.

Every aspect of the design of a building mechanical system relates to energy economics. Additionally, any and all other forms of energy consumed in a building ultimately decay to heat which deducts from the necessary heating energy required or adds to the cooling system energy. Thus, all disciplines of engineering in building systems impact the energy requirements. This observation is not unique to buildings; it can be extended to any system which consumes energy—transportation, industrial processes, etc.—which is to say, that those concepts which apply to the reduction of the waste of energy in building systems are directly applicable to energy engineering in automobiles, airplanes, pipelines, heat treatment, assembly process, and all other processes, devices, and systems that consume energy. The energy required to satisfy the product need must be defined, and every reasonable effort must be made to reduce the energy potential in the fossil fuel consumed until it is as close to the product need as the laws of physics and thermodynamics will allow.

The contents of this book are intended to assist the reader be he (or she) a student, engineer, manager, or legislator, in some small way, in enabling him to do his part in contributing to a solution to the energy dilemma.

Acknowledgments

The author wishes to acknowledge the assistance and cooperation of the original publishers of the materials in this book for their cooperation and encouragement. These publications are *Heating/Piping/Air Conditioning*, *ASHRAE Journal*, *ASHRAE Transactions*, *Purdue University* and *Specifying Engineer*. Recognition is also given to my colleagues Philip D. Sutherlin and Albert W. Black, III, coauthors of some of the original articles. The appendix contains a detailed list of the original publications, titles, and dates.

Special recognition is given to Robert W. Roose who provided counsel in the organization and planning and extensive assistance in the rigors of editing this work.

WILLIAM J. COAD

Contents

SECTION I

Engineering philosophy

Thomas A. Edison, recognized as one of mankind's most talented inventors, once said, "Genius is one percent inspiration and ninety-nine percent perspiration." In uttering this statement he failed to recognize another important ingredient. There are many brilliant technicians who have untold inspirations, and there are many who perspire a good deal—oftentimes the same individuals. But this combination of inspiration and perspiration does not necessarily result in the product of "genius." Continuing the premise, if one examines the field of engineers, from students to experienced practitioners, one finds that often the most brilliant or talented, from the standpoint of fundamental knowledge of the relevant engineering sciences, is not necessarily the one who displays the most productive creative ability, although all those considered may perspire an equivalent amount. The difference, the third ingredient, could be called the *necessary philosophy* or *attitude*.

This philosophy cannot be wrapped up in a simple small package and distributed to the engineering student. Psychologists may study its existence and debate whether its possession is a result of inheritance or environment. Many have often heard the expression "Engineers are a strange breed," a statement which, indeed, has a basis in fact. An engineer is as much a creative artist as a painter or a sculptor. A "good" engineer is, indeed a master artist. Using only the laws of physics and chemistry as his brush and canvas, and the material resources of the world as his paint, the engineer "creates" machines, systems, gadgets, and devices that become such a common part of one's existence that we scarcely notice them. It is as if Vincent van Gogh not only created his masterpieces but developed a machine to make a perfect reproduction available to all who desired one!

Most engineers may not consider themselves as creative artists, but if they are not creative, they are not fulfilling the most rewarding role of the engineering profession. Probably the most notable example of this analogy in history is Leonardo da Vinci (1452–1519) who indeed is considered one of the greatest artists *and* engineers of all time. This was not unusual during the Renaissance era. Although all engineers have not been gifted with the talent for painting or sculpture, all good engineers have developed the philosophy to use the tools at their disposal, as did da Vinci, to develop or create the most desirable or useful engineered product as the charge may be given him. This product could be a bridge, a thermodynamic machine, or a household device. The products of such engineering genius have done infinitely more to change the destiny and life-style of mankind than have the works of the greatest masters of the arts.

This section includes some discussions on various concepts of engineering philosophy. Hopefully, the reader can start piecing the puzzle together from the content of the chapters in this section. From these chapters one will recognize that to be a successful or creative engineering designer first requires a thorough knowledge of the relevant laws of physics and chemistry. Without this knowledge, the engineer is like the painter without a full inventory of paints. From this starting point, it may be that the pieces of the puzzle are endless (that is to say that anyone who held all the pieces would essentially possess the secret of life). The discussions, however, are intended to include some of the key pieces—certainly enough to point out the proper direction.

If the knowledge of the physical principles could be considered the foundation, the cornerstone would be curiosity. Curiosity not only for its own sake, but as it relates to the chore at hand. The good engineer is never satisfied with a lack of thorough understanding of the problem. If he encounters a problem which he does not understand, he will not simply rest comfortably with the "assurance" of a colleague. He will question, study, and research until he possesses the understanding and the knowledge that he is comfortable in this understanding.

But the building is not constructed of a foundation and a cornerstone alone. And the point of this discussion is to illustrate that the foundation does not a building make! Other component parts, properly placed are necessary.

The chapters in this section start with a discussion on the difference between pure and applied science. The reason this topic was authored is that many engineering efforts have become deadlocked in the arguments of thorough scientific understanding. From this introduction, several other key ingredients are presented such as simplicity, the use (or misuse) of so-called rules of thumb, how neglect of minor details leads to design failures, with the final chapter in this section addressing the economic question—can our economic system afford the price of engineering?

The chapter on the "New Energy Technology" is intended as a lesson to the engineering practitioner to recognize the true components of the "big picture," but also presents a message to those who would look for greener pastures across the fence, when the finest fodder is beneath their feet.

1

Pure versus applied science

Progress in engineering technology is in considerable danger of being thwarted at this time when the social impact of the rapid advances made in the preceding half century is mandating the need for progress. The dangers, strangely enough, are being created by the very sectors that are attempting to accelerate the progress.

Some such social impacts, for example, are those of energy consumption and environmental effects—both engineering problems created by engineering technology and, as such, deserving of solution by expanded engineering technology. The current trend, however, is to retard progress in achieving the solutions until the solutions can be addressed from the standpoint of pure science. This trend, in part, appears to result from a lack of understanding of the difference between *pure science* and *applied science*.

These two phrases are defined by Webster in the following manner:

• Pure: restricted to the abstract or theoretical aspects, as *pure* physics; contrasted with *applied*.

• Applied: used in actual practice or to work out practical problems, as *applied* science; distinguished from *pure*, *abstract* or *theoretical*.

The fundamental sciences, such as physics and chemistry, can be approached as either *pure* or *applied* sciences. Engineering, however, is not a fundamental science: it is by definition an *applied science*. Engineering utilizes laws, physical relationships, and other exacting knowledge developed within the pure sciences as its cornerstones. It then builds upon these with corollaries, hypotheses, principles, observed laws, and observations of

physical relationships to change the forms of nature to those desired by man. Each time such a change is accomplished, the original order of nature is affected. This, of course, is the cause of the depletion of the natural resources that are used to accomplish the changes and environmental transformations that are brought about.

As these effects compound to a magnitude implying retardation of continued advancements in technology or adverse influence upon society, the applied science must be expanded to address the newly recognized problems. The solutions cannot await the rigorous and cumbersome cornerstones of pure science. Compared to the pure science, applied science can respond most effectively and rapidly to generate solutions to problems and subsequently accomplish the needed change. The pure scientist *must* deal in exacting terms: the number of electrons in a molecule, for example, is an exacting statement. The applied scientist, on the other hand, can function quite comfortably with such phenomena as *order of magnitude* or *significant figures*. Had mankind historically awaited the assurances of perfect knowledge offered by the pure scientist, it would never have had the first bridge, controlled combustion, or the wheel.

As the history of applied technology has advanced, the inevitable refinement of the orders of magnitude and assimilation of hypotheses, observations, and the like, have created the exponential curve of progress. For some, this refinement of the orders of magnitude and the extensive bank of technical development appears to have created an element of confusion between exactness and purity. That our electronic data processing tools can provide the answers to complex engineering problems to "n" significant figures does not necessarily assure any more reliability in the order of magnitude than the understanding of physical phenomena or algorithms which define the problems.

Conversely, if the algorithms, when properly applied, provide an answer that successfully produces the information to enable the engineer to accomplish the desired "change in the form of nature," they need not be discarded until verified by the techniques of pure science. *This is the essence of engineering*!

This discussion relates to all fields of engineering, but a specific example of concern regarding building environmental systems technology relates to load calculations and calculations regarding anticipated energy and energy resource consumption.

Determine energy resourse use

The salient ingredient of the building energy consumption calculation is the time-integrated space load, including the quantitative effects of the load resulting from the external environment. This load response is a very complex phenomenon involving weather anticipation, thermal transfer characteristics, thermal storage characteristics, etc. In the field of building energy calculations, lack of agreement (and progress) on the *pure* method of calculating this integrated load has significantly retarded progress toward the ultimate recognition of acceptable methods of addressing the problem—*determination of the energy resource consumption*! The integrated load, although the salient ingredient, is but one input. Other aspects of the problem, such as system analysis, component performance, occupancy schedules, and operating modes, can often have a much greater impact on the consumption than the integrated space load.

Consider new approach

It is time for the engineering community to consider a new approach to our goals. We must accept the lack of purity, move on with what we have, and subsequently readdress the problem at a later time when (and if) purity has been achieved.

2

The engineering design process

All the disciplines of engineering can be subdivided into two basic categories: *design* and *research*. Research, in turn, can be divided into *pure* and *applied*, with applied research in fact oriented toward the solution of a design problem. Thus, the vast majority of engineering professionals or practitioners are engaged directly or indirectly in the design process.

The process of developing a design is an interesting one, whether the design be of a component, a subsystem, or a total system or product. The first observation is that there are two basic approaches to the philosophy of design of engineered systems. One is to start with an inventory of the known available products, define the parameters of the anticipated system, and then set about the task of accomplishing the resolution of the parameters with the products of the inventory to achieve the system. The second approach is to concentrate on the specific definition of all possible parameters for the system, subsystems, and components and from these develop an idealized design, then temper the final design by reducing the ideal to that which can be accomplished with the available products.

Designer's knowledge important

In the first approach, the design is limited by the designer's initial knowledge of the available components; whereas in the second approach, the ultimate design is limited only by the designer's ingenuity in defining the parameters, with the *total* available components simply defining the limits. In the second case, not only does the lack of knowledge of the available component products not restrict the results, but, because the designer achieves a preliminary result *prior* to inventorying the component products available, he has considerably better knowledge of what he is looking for when he does conduct the inventory. Furthermore, the designer who follows the second approach has historically been the catalyst of progressive inventions of new products.

Another similar observation that carries the *available product* foundation of the design process one step further is the *handbook* approach. This approach to design is to consult reference materials and publications to determine how past designers have developed previous similar designs, and then to use their concepts as though they were an available product.

Other ways of expressing this difference in design philosophy are:

• *Continuity of methods* versus *creativity.* This is the approach wherein the designer develops his new design utilizing the same fundamental parameters he employed in the past, rather than continually updating his definition of parameters and forcing them into a specific project.

• *Product orientation* versus *systems orientation.* In this case, the difference must be recognized between a final system product and a component product. As an example, the designer of an electrical resistor or transistor has completed the design (and subsequent manufacture) of a component product. It is the system designer who utilizes this product as a component of a television set, calculator, or whatever. Similarly, the designer of a

5

window air conditioner or a centrifugal water chiller has provided a component for a building environmental system. The component products are *not* the completed system until they are integrated into the systems in some planned or unplanned array of other devices. Lack of recognition of the difference between the product and the system inevitably leads to inferior ultimate performance.

When one observes the myriad of engineered products and systems over the years, it is evident that the vast majority were designed on the basis of the *available component product* or *prior experience* philosophy, rather than the *system* philosophy.

The design of the ultimate system or consumer product, when based on the first of the philosophies discussed, tends to fix the state of the art; whereas, when based on the second philosophy, it tends to advance the art. As one observes specifically the HVAC industry over the past three and one-half decades, the most striking example of the overabundant application of the product-oriented philosophy is the evidence of fads, such as radiant heat, dual duct high velocity, perimeter induction, absorption cooling, total energy, all electric, variable air volume, and so on. These fads varied more with calendar time than with any other recognizable parameter. Yet calendar time has little to do with the specific parameters relating to a specific building project.

System design requires talent

Admittedly, the second or conceptual approach to systems design may require more engineering talent and time. Talent being in short supply and time being an economic fact of life likely justifies having to use the first approach. However, we are at a stage in both the national and international economic cycles where we can no longer afford the luxury of waste caused by a static state of the art. The day of being successful in the automotive industry by making last year's product with different trim is past. The day of building the new glass-skinned skyscraper with a dual duct high velocity system has also passed.

Unlike the consumer product business of optics, electronics, and automobiles, the building industry is not motivated by foreign competition. However, as our economy becomes ever more sensitive to capital shortages, reduced growth of productivity, increased operating and maintenance costs, and energy shortages resulting in increased costs, the need for recognition of the value of the conceptual approach to designing engineered systems will inevitably surface.

3
Design parameters

Whether it be in operating a business, creating an artistic work, working at a skilled trade, or practicing engineering, there is an inherent tendency to filter out many of the seemingly insignificant bits of basic knowledge upon which our expertise was structured and base our daily functioning and decisions on selected conglomerations of these basics. This developmental process comes with experience and is a prerequisite to effective production.

From time to time, however, with changing technology, social structures, monetary economic conditions, etc., every practitioner should reevaluate the parameters upon which he is basing his daily decisions. The reevaluation of parameters, if applied to engineering designs, would produce vastly different results from those we have seen in the past two decades.

To develop this concept, consider that as any new field or engineering technology is born, there is essentially only one design parameter—performance. Examples: that the machine convert thermal energy to shaft work; that the generator convert shaft work to usable electric energy; or that the bridge span the river and support the weight required. Then, as each field of engineering technology matures, other design parameters inevitably are required. Take, for example, the heat engine, for which the original parameter was simply to convert energy from a thermal to a mechanical form. Subsequent parameters that evolved included improving the heat rate (or energy efficiency), decreasing the weight-to-horsepower ratio, reducing the maintenance requirements, refining or improving the automation and safety systems, and decreasing the production or manufacturing cost per unit of power produced.

As the practicing engineer embarks upon any phase of design, he is well advised to stop for a moment and compile a list of the design parameters that he will attempt to satisfy. In the field of building systems design, as in any discipline of systems engineering, such an examination of relevant parameters must be undertaken at numerous phases throughout the design development. A typical listing of such phases relating to building environmental systems would be:

- Establishment of indoor conditions.
- Calculation of the loads and load profiles.
- Selection and design of terminal control systems.
- Selection and design of terminal distribution systems and methods.
- Selection of type(s) of thermal distribution systems and subsequent design.
- Selection and design of ranges, rates, and thermal levels of thermal distribution systems.
- Selection and design of high-level (heating) primary conversion system.
- Selection and design of low-level (cooling) primary conversion system.
- Selection of high-level (heating) energy source or sources.
- Selection of low-level (cooling) energy source or sources.

Consider, for example, the first phase listed above—establishment of indoor conditions. If this question were addressed hastily regarding air conditioning for human comfort, one might state a specific dry bulb temperature and relative humidity (such as 75 F DB and 50 percent RH). But upon reflection, we all realize that the basic parameter is human comfort. Thermal comfort, in turn, is a physiological phenomenon achieved by a balance between metabolic heat rate (input), work produced (output), and heat dissipated (re-

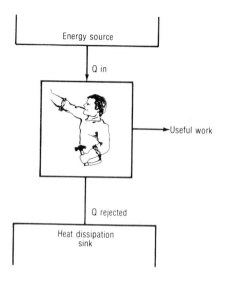

Energy source

Q in

Useful work

Q rejected

Heat dissipation
sink

Fig. 3-1.

jected). In oversimplification, the human machine can be described thermally by the classical block diagram used in elementary thermodynamics to illustrate the second law (see Fig. 3-1). The designer of the thermal environment is concerned chiefly with the heat dissipation. Consideration of the heat transfer phenomena affecting this exchange—radiation, convection, evaporation, and conduction—establishes the following comfort parameters:

- Dry bulb temperature.
- Relative humidity.

- Mean radiant temperatures (enclosed and "exposed" surfaces).
- External radiant effects (solar, direct and reflected.)
- Contact surface temperatures (chairs, desks, etc.).
- Air velocity.

Obviously, this is quite an expansion from simply dry bulb temperature and relative humidity. However, thousands of buildings enclosing millions of square feet have been constructed without the ability to provide for thermal comfort because of a failure to recognize one or more of these as relevant parameters. This chapter does not include a discussion of some of these typical failures, but every practicing engineer can provide his own.

Another interesting aspect: It becomes immediately evident that the thermal environmental system is not *added* to the building; rather, it is an integral part thereof. The design engineer, therefore, cannot avoid involvement in the basic building design, or at least in the aspects related to heat exchange.

Other chapters will deal with parameters relating to the other phases of design. In the meantime, compile your own listings; in addition to improving the wisdom of design decisions, this will make it increasingly evident why the engineer must participate in the architectural aspects of building design decisions.

4
Evaluation functions

An evaluation function, as used in the engineering design process, can be defined as a relationship of one characteristic of a system to another that can be utilized in some way to evaluate the design, performance, or other similar merit of the system. Whether or not one has ever consciously considered it, an evaluation function is one of the most powerful tools in achieving successful engineering designs with a minimum of error. Unfortunately, some engineering practitioners become so convinced of the overwhelming value of evaluation functions that they tend to totally replace the normal process of engineering analysis with the evaluation function, basing a design on the function alone. Misused in this way, the evaluation function becomes what is commonly called a rule of thumb.

Properly used, on the other hand, an evaluation function can be employed in the early stages of the design process to set budgets or goals—in this way, it is used as a self-imposed design parameter—and in the intermediate and final stages to "check" both manual and computerized calculations. And finally, it can be used to determine many aspects of the success of the design.

Consider some examples of evaluation functions, categorized by general areas of relevance, with some commentary as to where in the design process they are employed.

Design capacity and loads

The function of space heating or cooling load per unit area of building can and should be employed as a design target or budget when the designs of the building envelope and interior energy dissipation systems are being developed. The commonly used units are power per unit area for heating (Btuh per sq ft) and area per unit power for cooling (sq ft per

ton). The "reference" values will vary depending on primary building use and location, but typical values used in the Midwest at about 40 deg north latitude are 30 Btuh per sq ft for heating and 400 sq ft per ton for cooling. Once the building envelope and internal energy systems (such as lighting) are established and the space loads carefully and accurately calculated, the specific load should be calculated to determine how closely the actual design compares with the "target" or the value considered the norm. At this point, not only is the success of the designs of the building envelope and internal energy systems evaluated by the function, but errors in calculation will also be flagged if a value varies significantly from the reference value. This is the "checking" use of the evaluation function.

Other capacity-related functions are air flow rate per unit area (cfm per sq ft) air flow rate per unit cooling capacity (cfm per ton), design water circulation rate in a chilled water system per unit installed capacity (gpm per ton), and specific water flow rate for a water heating system (gpm per MBtuh). A brief look at two of these reveals some common misuse of evaluation functions.

Air flow rate per unit area is quantitatively related to a more fundamental design parameter, air changes. In years past when the air was being "changed" in a space for some purpose other than maintaining comfort conditions with mechanical cooling and heating, air change rates were used as an absolute design tool. More recently, minimum air circulation rates were found to be necessary to achieve adequate ambient velocity for convective and evaporative cooling comfort. Subsequent study of air distribution devices, however, has revealed that previously established minimum air circulation rates are not absolute values but rather are related to the location, shape, and design of the supply air device. Yet there are building codes in some areas that actually require minimum air change rates or minimum cfm per sq ft values.

Another example of a common misuse relates to the water flow rate per unit installed capacity for a chilled water system. The gpm per ton is a useful evaluation function properly applied. It should not be used, however, as a design value in designing the system. The proper approach should be to design each load device (coil) so as to achieve the most beneficial resolution of air temperatures, flow rates, and pressure drops; water temperatures, flow rates, and pressure drops; and cost burdens of the coil and air handling unit. It is then that the system flow rate is calculated, and the gpm per ton function reveals the success of the work by comparing it to a norm. If a norm value of 2 gpm per ton is used and the answer comes out to 1.2 gpm per ton, this would indicate an extremely successful design from the standpoints of pumping energy and piping system costs. If the answer comes out to 3 gpm per ton, a reconsideration of the design should be undertaken.

Power and energy consumption

A brief summary of some of the more common functions relating to power and energy are annual thermal energy per unit area of building (Btu per sq ft per yr), annual electric energy per unit area (kW-hr per sq ft per yr); electric power requirements of cooling system auxiliaries per unit installed capacity (kW per ton), annual electric energy requirements of cooling system auxiliaries per unit of refrigeration produced (kW-hr per ton-hr), electric power requirements of refrigeration machine drive (compressor) per unit capacity (kW per ton), and annual electric energy requirements of refrigeration machine drive per unit of refrigeration produced (kW-hr per ton-hr).

These functions have become commonly used in energy management efforts and are being extended into the design development process.

Economic evaluation functions

The commonest economic evaluation functions, familiar to all designers, are system cost per unit area of building ($ per sq ft) and system cost per unit system capacity (such as $ per ton).

There are other functions that are actually building block components of these two, such as $ per cfm for air distribution systems and air handling units. In addition to their use in evaluating the economic success of a design, these functions are commonly used in the development of preliminary cost estimates. Another interesting function that can be used in the development of preliminary estimates is the weight of sheet metal in the distribution duct system per unit volume of air delivered (lb per cfm).

In building management, the function of energy cost per square foot has been used historically. Recent studies have revealed a close interrelationship between energy consumption and maintenance service costs. It has been found that the sum of these two (energy plus M/S) in the numerator often provides a more meaningful function. In revenue-generating building projects, the dimensionless function of the energy plus M/S costs divided by the gross revenue has proven quite useful.

Other evaluation functions

Another type of evaluation function not addressed above (except the last one) includes dimensionless ratios, which have the same uses as the dimensional functions discussed. Dimensionless functions have additionally been found to be valuable tools in mathematical formulations of physical phenomena. Examples of these are expressions of efficiency, coefficients of performance, and the dimensionless numbers such as Reynolds, Prandlt, etc.

There are many other evaluation functions, including those that are commonly used and those that individual designers have formulated and developed over the years for their own use. The vast majority of those in the latter category have never been printed or even written down; they are simply the mental tools that have enabled their users to develop successful designs. A major component of the "experience" that enables designers to improve with time is the continual development of a growing reservoir of evaluation functions that are employed consciously or subconsciously.

5

The influence of "fads" on design

When one hears talk of fads or something that is faddish, one generally thinks of such things as clothing styles, music, various art forms, or the like. And if such philosophically debatable aspects as the social impact of music, art, or clothing are overlooked for simplicity, fads in these areas can be considered harmless and not detrimental to the social welfare.

When fads leave the arts and enter the sciences, however, they can be generally detrimental to the social welfare for the simple reason that reliance upon a fad as the status quo or state of the art will inevitably hamper progress toward better solutions to scientific problems. Scientific practitioners lean toward the comfortable position of sharpening their skills on the current fad in preference to the less comfortable posture of seeking better solutions. The latter situation is sometimes considered as a socially undesirable position of "opposing the trend."

Are engineering fads common?

To develop the hypothesis, consider first the question: Are fads the common mode or state in engineering practice? The rhetorical answer is *yes*, with a few simple examples to justify it.

A combination of marketplace forces have built faddism into engineered products, whether they be consumer products or less visible products such as production tools and structural shapes.

Probably the most obvious consumer product is the automobile. Engineered designs in the United States followed very distinct fads, commencing, say, in the mid 1940s. Looking back, we see a progression of fads from lower to longer to wider to more chrome to higher horsepower. In this progression, virtually all

engineering talent was directed at satisfying the then-current fad.

Less evident to the general public, but certainly known to practitioners in the building business, (both those in the technical end and those in the financial end) is the faddism in architecture. Perhaps because of the relationship between the disciplines of art and architecture, it can be understood how the susceptibility to fads spills over. From the standpoint of financial stability, architectural fads can be either an asset or a liability. A short-lived fad that quickly "ages" a building can be disastrous to the investor, whereas a long-lived fad can stabilize the investment.

Turning to a less visible component of the building, the mechanical system or the environmental system, fads have been most prevalent. In the late 1940s, hot water radiant panel heating systems were faddish. They were dutifully applied and misapplied by designers of systems for office buildings, schools, churches, homes, and virtually every other conceivable type of building. Rather than recognizing that the major problem with these systems in their areas of misapplication was one of thermal mass creating a system time constant that could not possibly respond to load changes, major control manufacturers encouraged their continued use and attempted to solve the problem by sophisticated anticipatory load controls.

As the building fads turned rapidly from double-loaded corridor shapes to cubical shapes with sealed fenestration, the need for total environmental control arose. The attendant need for multiple zones of control combined with architectural pressures for "volumetric efficiency" and the coincidence of inexpensive energy brought about the fad of the high-velocity double duct system. This has been the most successful and long-lived fad since the steam radiator. Even in the days of "cheap" energy, if thorough system analyses had been done, a major percentage of the high-velocity double duct systems could not have been economically justified.

What forces motivate faddism?

Before discussing current fads, a second question should be addressed: What force or forces in society motivate the engineering community in the direction of faddism? The answer to this question is not simple since there are many interacting forces at play. A simple statement of the major contributions might be:

• *Economics of mass production.* From the standpoint of the manufacturer, mass production is the name of the game in cost reduction. Cost reduction gains the favorable position in the marketplace. And mass production requires large volumes of items in standard shapes and sizes. Thus, the manufacturer who tools up to produce a product (say a high-velocity double duct mixing box) must promote a market to assure a volume sufficient to control the price that the promoted market can endure.

• *Economics of risk.* Once a market is assured (i.e., the fad is established), the safest venture on the part of any manufacturer is to provide a product that is usable within the established market. This is the "sure thing" approach. The example stated above of the control manufacturers who developed products to improve the performance of misapplied radiant panel systems is but one of many such examples.

• *Economics of cascading opportunity.* This is the catalyst that generally motivates success in both of the above cases. In the vast majority of situations, particularly in the building systems industry, the conceptual "system" required products of several different component type manufacturers. As an example, with the high-velocity double duct system,

the mixing box was only one small component. Along with it was needed large, high pressure (Class III) fan equipment, high-pressure duct apparatus, special vibration dampening devices, and acoustic attenuators in the duct systems. Each of these devices opened a gate to another group of product manufacturers, all of whom subsequently supported the concept! This cascading opportunity phenomenon carried on in many cases to the energy supplier who sold the energy to motivate the system. In numerous fad examples, the energy supplier became both the prime beneficiary and the prime promoter of the fad.

With the existence of fads in the engineering areas of building environmental systems recognized and the forces causing these identified, we might now address current fads. If the multitude of current fads could be expressed in one word, it would be *energy*. Some rather evident, capsulated statements of these faddish systems and devices are: variable air volume, variable inlet vanes (fans), variable speed pumps, heat wheels and other heat recovery devices, central digital computerized control systems, and a host of other products to save energy.

Faddism is not the best solution

At the outset, it was stated that ". . . reliance upon a fad . . . will inevitably hamper progress toward better solutions to scientific problems." The development of this hypothesis endeavored to illustrate that the establishment of fads is necessary to assure some element of stability in the manufacturing of engineered products. But nowhere in the hypothesis is it stated that the best solution to any given engineering problem is provided by the reliance on or acceptance of a faddish product or system! Thus, if the consumer in the area of building environmental systems is to realize the best and most favorable economic and performance solution to his needs, it will not lie in fads but rather in both system and product engineering tailored to the needs of his specific situation.

6

The value of simplicity in design

If you can't draw a picture of it, you don't understand it.

If one rule relating to success in original engineering design efforts could be said to be the *most important* it would be: "keep it simple." Two fundamental observations leading to this conclusion are:

- Inspection of numerous systems, say, five years after installation and start-up reveals that the vast majority are not being operated as intended by the designer. In many instances, this abortion of the designer's intent has resulted in less than satisfactory performance. As the investigation proceeds, it is found that the agent responsible for the operation has simply reduced the complexity of the system to his level of understanding.
- In the field of original engineering design, many of the concepts incorporated are a result of deficiencies revealed by previous designs that are subsequently resolved by *adding* a correction. As these corrections compound, system generation after system generation, they come to be accepted as the norm.

When a system designer sets out to define the parameters that he is attempting to satisfy by the design, each step in the process should be aimed at not only satisfying the parameters but also at satisfying them in such a manner that he fully understands how the system dynamics are going to respond to each operating mode and statepoint dictated by the load. Once the problem is addressed in this manner, the designer will inevitably achieve a high level of simplicity (not a contradiction).

Consider this example

As an example, consider the design of a hot and/or chilled water system—a rather common subsystem in buildings or building complexes.

These are relatively simple in concept, consisting of a generator, expansion tank, circulator, load, and piping system. However, as designers attempt to satisfy various reduced load conditions and multiple modes of operation, such devices as load control valves, three-way mixing valves, secondary pumping, multiple expansion tanks, emergency system interconnections, system bypass valves (for constant chilled water flow), etc., are added, not to mention the inherent complexity resulting from sheer size and scope. As a result of this, there are an untold number of systems serving large building facilities and campuses today that have become so complex that thermal-hydraulic network analyses are virtually impossible to perform.

This degree of complexity must be avoided with deliberation by the designer. Although at first glance the response might be that simplicity cannot be achieved and the multitude of performance parameters still satisfied. There are, however, some rather fundamental rules that can be followed that will assist the designer in accomplishing this goal. For hydronic systems, these rules might be as follows.

Hydronic systems design rules

• If two or more systems are hydraulically interconnected (such as with campus systems or two-pipe heating/cooling systems), *never* have more than one expansion tank. The expansion tank, in addition to serving as an expansion volume for the liquid, is *the* constant pressure point of the system. If more than one such point is attempted, control of the system pressure is virtually impossible to predict under an infinite number of different dynamic response load conditions.

• When multiple pumping circuits are interconnected, such as with primary-secondary pumping, design for dynamic response isolation between the pumping systems. The secondary pump performance, for instance, should be virtually independent of the primary pump. With a little study, this independence can be achieved quite easily in most designs by the rigorous application of the "common pipe" principle.

• Try to achieve the design without the use of three-way mixing valves. Most control response requirements can be achieved with globe or "two-way" control valves. When three-way valves are used, it is easy to fall into the trap of unanticipated series pumping. Furthermore, the part load hydraulic performance of three-way valves is, to say the least, ill defined in practice.

• Consider secondary pumping of source devices such as boilers and chillers. Properly designed, this can be accomplished as economically as such alternatives as load bypass valves, chiller two-way valves, or load three-way valves, with more predictable results and reduced part load energy consumption.

• Draw a flow diagram of the entire system at the early stages of design and again after the design is complete. There is a philosophy in addressing technical problems: if you can't draw a picture of it, you don't understand it. The reverse is also generally true. The flow diagram (preferably in ladder diagram form) is a comprehensive picture or schematic of the thermal-hydraulic system. It enables the designer to achieve understandable simplicity and readily identify designed-in problems, serves as an aid to the installing agents in interpreting construction diagrams, and provides an invaluable tool in the ongoing operation and diagnostic servicing of system problems. The flow diagram, well developed, will immediately reveal the design philosophy of the system.

The hydronic system has been used as an example to illustrate the concept of designing for simplicity. This concept can and should be extended into each and every subsystem and component. If done successfully, system performance will be improved, and as a bonus benefit first cost will generally be reduced.

7
Specialty devices in engineered systems

In the integrated building environmental system, all major subsystems contain, in addition to the fundamental components, a number of devices, generally with moving parts, called specialty devices. The fundamental components are considered those that theoretically would function as a workable system under ideal conditions with steady-state capacity requirements. Examples of fundamental components are:

- *Refrigeration system* (vapor compression): evaporator, compressor, condenser, throttling device, and the three major piping segments.
- *Hydronic system:* source, load, circulator, compression chamber, and interconnecting piping.
- *Steam system:* source (generator), load (condenser), steam piping, and return piping.

Identify subsystem components

Such a grouping of fundamental components can and should be identified for each major subsystem.

The specialty devices are all the other devices that, rightly or wrongly, are employed to make the system function. These devices fall into one or more of three categories:

- Safety devices or controls
- Devices necessary for normal operation
- Indicating or status information devices

The importance of understanding the significance of these devices cannot be overemphasized. The simplest of systems are those that contain *no* components in addition to the fundamental components. The addition of any of the specialty devices in the second category above adds a degree of complexity to the system. Furthermore, the addition of any specialty device, assuming it is absolutely necessary to the safe and proper operation of the system, is one more element to maintain and service throughout the life of the system. If it does not warrant such maintenance and service attention, it should not be employed!

Many of the devices in all three categories have been developed in reaction to problems encountered in system operations. In many cases, rather than "designing out" the need by improved systems designs, the design community turned to designing the systems around the available specialty devices, thereby promulgating both the need for and the dependence on such devices. This dependence has influenced both system designers and product designers more than has fundamental design philosophy.

Reliability is vital

In addition to the disadvantages of complexity and maintenance service burdens, another significant penalty for the excessive use of specialty devices is the introduction of unreliability. The vast majority of such devices contain moving parts. Whether these parts be bearing surfaces (rotational or sliding), metal flexing (bourdon tubes or expansion compensators), or elastomer diaphragms (pneumatic relays), they are subject to failure! Such devices therefore lessen the anticipated statistical reliability of the system. To add to the reliability problem, the designers of specialty devices have, in many cases, value engineered the devices to an extremely low level of reliability to gain a more favorable position in the competitive marketplace. The subsequent acceptance of such lower cost (less reliable) products by the consumer (the system designer) has forced the entire market in the direction of less costly, less reliable devices. We thus find ourselves in a situation wherein the state of the art in systems design is dependent on myriad devices that in many cases, because of competitive pressures, are not available in high quality or high reliability products—at any reasonable cost!

As a brief example, take the case of air elimination in hydronic systems. The *fundamental components* concept assumes that the only fluid in the system is water. In reality, the available water contains dissolved gases that release to the gaseous phase at elevated temperatures or reduced pressures. This "air," when separated, causes restrictions of liquid flow at high points in piping systems, as well as corrosion and other undesirable conditions. Designers have been provided total flexibility of piping system design by the availability of automatic air vents that have been employed at all geometric high points in the system, as well as at the points where the pipes must drop to avoid interferences. Such devices sometimes fail open, releasing water from the system, which results in vapor binding and subsequent circulation failure in the whole system. The total failure problem can be solved with another specialty device, the automatic (usually pressure-controlled) water makeup valve (or open connection pump). This device enables the system to continue operating even though the "vent" is malfunctioning. Inevitably, however, the resulting continuous makeup water, containing calcium that fouls the high-temperature heat exchange surfaces, causes extended shutdown for descaling.

The obvious solution is for the designer to design the system so that air can be eliminated from the geometric high points at the time of filling the system, and so that there are no other gas collection points. He must design a single, high-quality, reliable air separation and elimination system that is easily accessible for regular monitoring, maintenance, and service. Reduced to a single point or device, a high-quality device can be employed at any reasonable cost.

Numerous similar examples could be given. However, the system designer might consider the general rule that the system should contain as few specialty devices as absolutely necessary, and those that are employed should be the highest quality, most reliable devices available.

8
No details are minor

The fundamental ingredient of a successful system design is the recognition of all parameters relating to the system to be designed. Assuming these parameters are properly and successfully identified and satisfied, the complete success of the design then depends upon what might be considered a second-order requirement; *attention to design details*. It is at this stage in the development that the designer must decide upon the extent of the detail included in the construction documents vis-a-vis the skills of the installing mechanics or contractor. It is likely that as many operating deficiencies in mechanical systems have resulted from an overly optimistic judgment of these available skills as from any other singular reason.

This is not to criticize the level of skills of the installing mechanics, but rather to recognize that as the machinery and systems become more and more complex, the design profession must assume an ever-increasing responsibility to address the construction details.

Simple examples cited

One simple example of such detail is the method of installing a thermometer well in a pipe. It is not unusual to find wells two inches long installed in a coupling that is well in excess of that length, thus holding the sensing insert totally out of the fluid system (the same is true of temperature controller wells). The error introduced by this detail may be insignificant where there is a considerable temperature difference between the fluid being sensed and the surroundings, such as with a heating water system at 220 F. But as the temperature differential decreases, as is the case for chilled water or condenser water systems, the error becomes most significant!

Other common problems found in the method of installation of such wells are thermometers located such that they cannot be read or wells in a position such that they cannot hold the heat-conducting fluids.

Another example of such detail is the method of connecting condensate drain lines from cooling coil drain pans. The lack of attention to this seemingly unimportant detail has been responsible for extensive damage to countless ceilings, sometimes presenting an unsolvable problem after the fact if adequate space was not provided in the original construction for proper trapping of the air flows through the piping.

If the interior of the condensate drain pan were at the same pressure as the drain outlet, there would, of course, be no problem—it would simply be a matter of gravity drainage of the water. However, since the drain pan is generally at a pressure greater or less than the drain line outlet, the pressure differential motivates a flow of air through the open pipe. The fluid dynamics relating to the effect of this air flow (in either direction) on condensate drainage from the pan and through the piping system is so complex that its results cannot in most cases be anticipated. To reduce this complex problem to a manageable level of understanding, it is common practice to provide a water seal trap simply to stop the air flow. The design problem, then, reduces to the proper design of the trap.

Design drain trap properly

Probably the most troublesome drain deficiencies have been with draw-through type units, where the condensate drain pan is on the fan inlet side and is thus at a negative pressure with respect to the surroundings. Figure 8-1

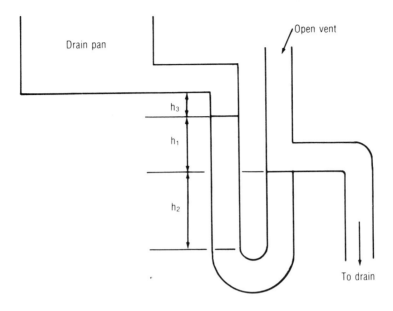

Fig. 8-1. Draw-through unit.

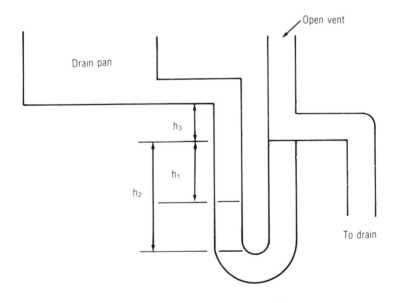

Fig. 8-2. Blow-through unit.

illustrates the proper trap arrangement for this type unit, shown in the operating condition. The trap functions like a simple manometer. It should be designed such that under no conditions is there any accumulation of water in the drain pan. The level of water in the right leg is, of course, established by the connection point of the outlet pipe. The difference between this level and the operating level of the water in the left leg is equal to the negative static pressure (in. WG) at the fan inlet. Dimension h_3 is simply a margin of safety which the designer establishes, depending upon conditions such as available space and accuracy of anticipated pressure calculations. The depth of the trap below the outlet (h_2) is critical. Upon unit shutdown, the level in the leg will drop to that at the outlet. Then, upon restarting, it will rise to the midpoint between the outlet level and the operating level. The water to provide this rise comes from the right leg, thus establishing the minimum dimension for h_2 as one-half h_1. To provide a margin of safety and to allow for evaporation, the recommended minimum dimension for h_2 is that equal to h_1. The open tee at the top of the right leg is necessary to prevent siphoning of the trap on shutdown, and it also serves as a point for priming the trap.

It is readily seen that such trapping can require considerable vertical space. If the drain pan is operating at 3 in. negative static pressure, the minimum distance between the bottom of the pan and the bottom of the trap is 6 in. plus the diameter of the pipe—generally not available if not planned for!

A similar logic applies to the trap design for the blow-through unit, wherein the drain pan is at a positive static pressure. This case is shown in Fig. 8-2, illustrating the conditions with the unit running. Dimension h_1 is the operating static pressure on the pan. The depth of the trap (h_2) must then be adequate to provide for this depth plus a margin of safety; the recommended dimension for h_2 is twice h_1. The distance between the pan outlet and the trap outlet (h_3) is, in this case, not critical if it is held at any value equal to or greater than zero. If, however, it is negative (i.e., the trap outlet is above the pan outlet), the difference in elevation should not exceed one-half h_1. If this were to happen, some of the operating volume of the trap would be lost on shutdown.

9

The new technology may be closer than we realize

It is sometimes alarming to observe the activities of the United States federal government in attempting to address the energy problem. Whether or not there truly is a problem is no longer an issue worth debate. The only question remaining is whether the people of the United States will learn to cope with the realities of energy economics. Failure to do so will almost surely lead to declines in our influence on the state of the world and eventually in our way of life.

The distressing thing is that our leadership appears to have its collective head in the sand—unable to face the realities of the situation.

A headline in a leading Midwest newspaper recently proclaimed: "Energy Plan: Re-Invent Car or Fuel." The news story, not too unlike many in recent days, included a report of the United States Department of Transportation secretary's description of a campaign "to reinvent the car" that will likely cost American taxpayers $500 million a year!

The fact is that the odds are very slim that throwing money at the energy problem, in the hope of uncovering a "new technology," will prove fruitful in time to prevent the first manifestations of the ultimate collapse. Realistically, the United States must start immediately on the long road to reduced energy consumption. Further, the only way to do this is to curtail use while putting our emphasis on applying *known* technology to increase the effectiveness of utilization.

There have been few if any breakthroughs in the fundamentals of energy conversion, except for nuclear conversion, since the nineteenth century. We have simply pushed the various thermodynamic cycles to the brink with respect to materials and safety; the same physical laws govern. Huge sums of money earmarked for new technology have been spent on the reincarnation of age-old concepts without thorough study of why the concepts failed to provide our energy needs during any of their earlier lives. Examples of these rediscovered "new technologies" include solar energy utilization, the heat pump, and integrated energy cycles (power and heat) and a move to a newly discovered savior referred to as "second law concepts."

The fact is that the "breakthrough" or "new technology" may be much closer than one would think. Further, the opportunities at conservation may be so significant that they could have a serious detrimental impact on our major energy-producing industries. If this is so, this is the area in which the government should be looking to solve a serious economic problem.

Energy conversion falls short

Using the second law as we have always used it, together with the Carnot analyses of efficiencies and $COPs$, it is immediately evident that today's articles of commence fall far short of achieving the ideal effectiveness in energy conversion. If efforts are concentrated here, trying to get closer to the ideal, the first effective steps will have been taken. This requires no new technologies, simply redirected efforts at heat transfer, fluids, and system parasitic analyses. In a building system, except for lighting, the end result of the entire energy system is to heat and cool. Heating is virtually a total degradation of useful energy to entropy. Entropy, in turn, is the end of the line in the conversion of energy resources to a useful form; this is the reason for the current trend toward heating with energy that has previously been used for a higher order form.

But to accomplish heat transfer in real machinery requires a finite temperature differential. Therefore, in considering cycles and machines that lend themselves to combined cycle application (power and *useful* heat), machines should be sought that deliver shaft energy efficiently while providing useful quantities of heat at elevated temperature. Such a device is *not* the external combustion Rankine cycle but rather the internal combustion Otto or Diesel cycle. Hardware is currently available that can be readily used in such applications; in fact, many such systems have been in operation for many years!

Consider the building system parasites. The purpose of the system is to convey heat into or out of a building. The major energy consuming devices in many buildings, however, are the fans and pumps that move fluids around—not the boilers or refrigeration compressors. To address this problem simply requires properly directed attention, not a new technology.

Consider thermal storage system

Current activity in solar research and application has produced almost an entire generation of "specialists" in so-called solar technology.

It must be recognized, however, that the collection of the heat is simply one small component of a necessarily complex system. In any thermal fluid system, when the need is not coincident with the availability, a complex system of storage is required. Efforts at advancing developments in the mundane concepts of hydronics would produce immediately beneficial results that could accelerate the feasibility of active solar systems. An example of lack of attention to "simple" technologies is the hydronic expansion tank. To obtain maximum benefit per investment dollar from a thermal storage system, the temperature range of the stored fluid must be maximized. This, along with the relatively large volumes of storage, requires the use of enormously large expansion tanks. Although there are methods available for coping with this problem at more reasonable costs, the current literature used by the vast majority of designers still contains the tank sizing procedures developed for systems of a different type.

Thus, the new energy technology is closer at hand than is normally recognized. It is in the less romantic areas of complex engineering technology, in controls, hydraulics, heat transfer, and fundamental thermodynamics. The problem is simply that these opportunities have been almost totally ignored in our efforts to find a solution that can be utilized with less need for basic understanding.

10

A re-examination of engineering education

Throughout modern history, one of the most written-about topics has been education, much of the writing done by the great philosophers and educators themselves. It is not the purpose of this chapter to enter the arena of the general philosophy of education, but some of this philosophy must be recognized in addressing the specific problem of engineering education and its capability to meet the needs of the profession and of society today.

H. G. Wells observed in *The Outline of History* in 1920, "Human history becomes more and more a race between education and catastrophe." With this in mind, we must reexamine our success in keeping ahead of the catastrophes as we see signs of them approaching.

Scientific and engineering education has enabled mankind to ward off catastrophes of many sorts. But in achieving this, we have created a world society that is heavily dependent on the resulting technology. A key element of this technology is a dependence on two very fundamental ingredients. One ingredient, limited by nature in its supply—which is rapidly diminishing—is fossil fuel. Another ingredient, limited by human resources and our economic system, is capital. As we see the possibility of a catastrophe resulting from the shortages of these basic commodities, we must look to education to gain a bit in the race to keep ahead.

An examination of the trends in engineering education reveals that perhaps there is good cause to initiate some improvements at this time. Engineering can be considered as a practical skill, i.e., the engineering practitioner generally applies the law of physics, channeled by guidelines developed through years of past experience, toward the end of producing a machine, structure, or other device to aid the productivity and comfort of man. This *channeling* effect has tended to steer engineering curricula into the path of *training*, rather than education. This trend has been strongly encouraged by demands of the marketplace that were a result of a total lack of comprehension or communication between the academic community and the business community. When the trend in engineering education was moving in the direction of producing theoreticians, the marketplace complained and put forth the cry for a more practical approach. This, it was felt, would make the engineering graduate more useful to the businessman. Without being recognized as such, the problem was one of pure versus applied science (Chapter 1); it was mistakenly identified as one of *education* versus *training*.

Educate rather than train!

The result is that the trend in engineering education has been to "train" engineers, rather than to educate them. Herein lies the essence of the problem!

In virtually any engineering curriculum, a certain amount of training is required. But, when the training becomes an end in itself, education has not been achieved and the curriculum has failed. The training is necessary to bring the student to the plateau from which the education can commence—it is only in this respect that education and training are synonymous. The training required preceding the process of total education in engineering is that portion of the curriculum that enables the educated engineer not only to understand and to make use of the fundamental sciences of

23

physics and mathematics, but also to have a comprehension of the current state of the art. The state of the art has grown and become so extensive during the middle half of the twentieth century that the problem of even fundamental training or skill development has consumed an overwhelming amount of the available time in the process of engineering education. As a result, the vast majority of curricula have lost sight of the goals of education and simply concentrate on training.

The observation is validated by a scan of not only the curricula of many undergraduate engineering schools but also the more popular textbooks. The trend has been to the "how-to" rather than the inquisitive "why-to."

The result of this has been the ability to develop, with the "how-to" state of the art knowledge, larger and more powerful automobiles, ever larger and more extensive building systems achieving the nth degree of the number of controlled zones and temperatures-humidity limitations, unprecedented electrical distribution networks, and countless other examples. With good fundamental training, successful design can be accomplished with a minimum of errors, but this also leads to very slow advancement in the state of the art. In some areas of the state of the art, advancement unfortunately ceases altogether where it is not pushed forward by immediate pressures from the marketplace—steam thermal systems, for example. Another disadvantage of a training-oriented curriculum is the failure of the practitioner to identify new parameters of design.

The leaders of engineering education should regroup, condense the period of training, and put more emphasis on education. The proposition is that *the ideal engineering education achieves adequate training, a rounded sampling of the humanities and related social sciences, and the stimulation of inquisitiveness and challenge*. It is the last of these that is needed so desperately if technology is to meet the challenge to ward off Wells' catastrophe.

To paraphrase Plato: "The direction in which education starts a society will determine its future."

11

Education: the primary ingredient in the solution of the energy dilemma

Few will deny that education is a fundamental ingredient in the solution of the vast majority of social problems. Ideally, an educated society is virtually devoid of ignorance. As such, it has an understanding of the interrelationships between the natural laws that dictate its course. Education instills in man's mind a curiosity that hungers for answers to those phenomena that are not readily understood. This curiosity generates explanations for what were previously considered mysteries, and so the cycle continues. An educated society, like an educated person, considers the unknown a challenge to be conquered rather than a threat to be feared.

Historically, totalitarian societies have succeeded temporarily by controlling all education systems and institutions. Such success has inevitably been short-lived, because to maintain the strength of the society, they had to continue to educate. Even though the content of the education programs was controlled, as the population became educated their curiosity and hunger for knowledge led them to question their society. This curiosity born of education thus becomes the first thrust to topple a totalitarian state's dominance.

Today, the world society is in the midst of an extremely complex and potentially dangerous problem—that of an accelerating consumption of nonreplenishable energy resources. Highly industrialized nations are totally dependent on enormous consumptions of these resources for economic survival. Densely populated nations depend on energy to maintain agricultural economies. Many

resources are owned by nations other than those who have a survival need for them. The world is thus close to sitting on a powder keg. The United States government has talked of such wishful solutions as "energy independence," a term coined to give the population peace of mind by thinking that we can get by with what we have. Three years later, however, we were importing more and domestically producing less than the years preceding "Project Independence."

How do these two observations tie together? The proposition is that if there is one key to the jigsaw puzzle of the so-called energy crisis, it is education!

The United States government currently is spending billions of dollars trying to cope with the energy dilemma, and the lack of success of these efforts is apparent to anyone who has been observing the continuing upward climb of the consumption curves. The reason is, stated bluntly, that when it comes to energy, the populace is ignorant. They can be easily misled by news releases, television commercials, and best-selling books written by self-proclaimed authorities, who offer impractical solutions.

Energy educated society needed

If it is accepted that the energy dilemma is a just national issue, then the federal government should channel a large portion of the funds appropriated for solutions into the educational field. We must become an energy educated society. These efforts should not be limited to those in the sciences, but open to all citizens in all areas of interest. If such a thrust were made today, within a very short period of time (say five years), the United States would be the world leader in energy management.

The proposed programs for energy education should address all concepts from the basic definitions, understanding where the energy to brighten the light bulb comes from, and a simple knowledge of how it was transformed from, say, coal into light, to the enormously complex problems of energy economics and sociopolitical implications.

Energy economics is a must

Energy economics should be taught in elementary schools, as required courses for all students in secondary schools, and as required courses for bachelor degrees in arts and sciences. For those specializing in sciences directly related to energy, such as mechanical engineering, physics, and geology, specific courses in energy economics as related to these disciplines should be required; and economics and political science programs should offer specific courses in energy economics.

Those who have already completed their formal education could avail themselves of the required knowledge through adult education programs, educational television, and other such vehicles.

The federal government could serve as a catalyst in this effort through such agencies and organizations as the Department of Education and the National Science Foundation. The programs could (and should) be tailored to fund the development of programs, courses, and textbooks, *not to provide them.*

A population with an understanding of the concepts of energy source and conversion, factual limitations, and impact of continued excesses would not find it necessary to turn to self-proclaimed leaders or to Washington for a solution to their inevitable problem, but would, with understanding, naturally develop their own solutions.

12

The professional's role in energy management

Design professionals must educate, motivate, and enlist others in the cause of energy conservation.

Engineers in private practice are considered professionals in somewhat the same manner that a physician or attorney is considered a professional. There is a difference, however, between the professional in private practice and his counterpart in a business structured toward the manufacture or sale of a product or other goods of commerce. This does not infer that this counterpart is any less ethical, moral, or dedicated.

It is helpful to develop a clear definition of the professional as he exists in the private practice of selling his services in the field in which he practices. One suggested definition of a professional in this context is: "A professional is a person whose primary obligation in all endeavors is to represent the interests of his client with integrity, and as rigorously as if they were his own."

If this brief definition is accepted as stated, it becomes evident that the problems of operating a professional "business" are different than those encountered in product-oriented ventures. Furthermore, no financial constraint is included in this definition. It might then be said, that as professionals, consulting engineers make or lose money when they negotiate the contract with the client. Once the contract is executed, the client's interests become paramount.

This chapter was reprinted from *Specifying Engineer*, April 1978 and was originally entitled, "The Professional's Role in Energy Management and Retrofit."

Why energy retrofit?

How does this relate to the topic of this chapter—the professional's responsibility in energy management? Why would a potential client want to undertake a retrofit program to reduce his energy or power consumption? The two fundamental reasons are cost and availability of energy. The last two decades have seen unprecedented growth in the building industry in the United States. Unfortunately, not only were energy and power requirements not considered as a design parameter during that time, but the prevailing economic forces were counterproductive to the concept of energy conservation.

Suddenly, previous, seemingly plentiful and inexpensive energy and power sources have taken an upward swing in cost, and in many instances, there have been curtailments in available supply.

Unfortunately, the American psychology finds it difficult to address this problem. This psychological difficulty is manifested in the "pass-through" cost concept, and more subtly in the cost-of-living index.

The pass-through cost problem is seen when one is operating a commercial building (leasing space to generate profits). If the energy costs go up, the tenant pays the excess. This same business scheme is utilized by product manufacturers, private or public institutions, and energy supply utilities (the latter in the form of "fuel adjustments" or "purchased gas adjustments").

The more subtle aspect applies at the level of the homemaker. When the cost of consumer products rises as a result of the multiplicity of energy cost pass-through—on every commodity from food to automobiles to widgets to utility energy—the cost of living is assumed to have gone up, and bargaining for a higher wage to cover the difference is initiated. Higher wages, in turn, stimulate another round of pass-through cost increases and the inflationary spiral soars upward.

Thus, in the vast majority of cases, the pure economic motivation does not stimulate adequate interest in energy conservation to lead a building owner to invest time, interest, or money in modifications or other efforts aimed at energy conservation.

Consider energy availability

Consider the other motivating force—availability of energy. Whether it be a production industry, an institution, a commercial building, or a homemaker—when the source of energy is curtailed or ceases to be available in needed quantities, the consumer reacts. This form of reaction can be described as panic, and the resulting corrective action is at best a meat-axe approach to a surgical need. Given a lack of proper understanding of the energy problem, when the temporary curtailment expires following adequate financial or energy source adjustments, the panic ceases and the user returns to business as usual until the next reaction is required.

As a result, there appears to be little reason why the potential client would want to undertake an energy retrofit program. The void, or lack of motivation, is filled either by the energy supplier who must stabilize his business to the consumer, or by the government (federal, state, or local) who feels the mandate to protect the public welfare, or by the product manufacturer who skillfully advertises "cure-all" products.

For the professional to simply react to the motivation stimulated by the energy suppliers, governmental agencies or product manufacturers, is to abort his first responsibility. Engineers are, or should be, totally aware of the complex nature of the energy problem and

should assume the leadership role in educating the public. If, as a profession, engineers had exercised this role of leadership in the previous two decades of astronomical growth in the energy-intensive building market, there would be less of an energy problem today.

The need for education

Educating the public, means everyone—including the professional's immediate client, others in the profession (both practitioners and students), product manufacturers, and even the homemaker.

Examine the implication of this concept. If engineers first accept the proposed definition of the professional and combine this "primary obligation" with a comprehensive understanding of energy conversion systems, then true "professionals" cannot refuse the role of leadership. How can design professionals cope with this overwhelming obligation to educate society when the vast majority have never considered themselves educators?

They must start by first educating themselves. There are reams of materials available in technical journals, daily newspapers, magazines, etc. There is no presupposing that all this information is valid or legitimate—contrarily, much of it comes close to propaganda, meant to lead the public or professional in a direction advantageous to the author's interests. Thus, this self-education must be undertaken with the attitude of a critic; the professional should challenge what he reads and generate his own beliefs. After embarking upon this exercise, most arrive at the conclusion that the energy and power requirements of our buildings and building systems have been extremely wasteful and excessive. Furthermore, regardless of whether the professional is personally convinced that ample supplies will be available in the immediate future, energy sources are depletable; that is, they are not infinite. Thus, any use today will deprive a future generation.

After reaching an understanding of this problem, the engineer can start directing his efforts toward the education of others. In the professional's design efforts, and in discussions with clients, it becomes easy to motivate his staff and clients in the belief of energy conservation. Most building managers and developers do not realize that the professional consultant or designer has substantial control over future operating costs.

To strengthen the professional's own understanding, and that of his peers, he should assume as active a role as his time will allow in the technical and professional societies such as ASHRAE, the Consulting Engineers Council, the Society of Professional Engineers, etc. Nowhere can he get so much and give so much with so little time devoted.

Community involvement may be more difficult, but the opportunities are endless and there is a place for just about any individual or personality. Examples include service clubs (such as Kiwanis, Rotary, Lions, etc.), local code committees, state energy advisory committees, guest lecturing in local schools, and working with building management and investor organizations.

Cost benefits should be goal

Some professionals become annoyed when they hear consulting engineer groups talk about "cashing in" on the energy retrofit market potential—competing with such forces as product manufacturers, energy suppliers, or "free assistance" offers by government agencies.

The fact is, *no* single product, device, or gadget can envelop the complex problem of energy conservation in *all* buildings, or even in a small sampling of them. Yet, virtually all energy conservation products have the ability to produce energy-conservation savings when properly applied.

The energy supplier attempting to simultaneously effect more productive use of his capital and labor investment and assist his client, informs his customer of those features of the available rates that will be mutually beneficial.

The governmental agency, usually adequately funded by some method of federal grant, offers tax incentives, a free-of-charge walk-through audit, or a handbook of energy conservation ideas.

But only the professional, who possesses the motivation and has the unbiased interest of the building owner or operator, coupled with a thorough understanding of the dynamics of building energy systems, can determine both objectively and quantitatively the true benefit of investing in product X, the modifications needed to take advantage of rate Y, or the actual significance of accepting governmental inducement Z. Only the professional has no motivating interest in either the sale of a product or the sale of energy. He is, or should be, intimately concerned with the cost-benefit implications of his recommendations and with client satisfaction with his involvement.

Energy management after audit

The question might be asked, "How do professionals inform potential clients of the benefits of employing professionals to direct their energy retrofit programs?"

Energy retrofit efforts are one small component of the overall problem. Furthermore, many successful energy conservation programs will require *no* retrofit at all! This is one of the determinations the professional must make.

The professional's entree into the project should be based on establishing an energy management program. Then, as with any effective management program, the first step is to establish the facts. This is accomplished through a detailed energy audit. This process entails identifying all operable energy systems in the facility, evaluating what each has contributed to the total energy demand (power), and determining what portion of the total consumption is assignable to each. (See Section IV.)

After developing these profiles and components of consumption (which must be verified for accuracy by comparison with energy bills), the professional can start on the sorting program. The sorting program separates energy consumptions into various categories—product energy, control energy, parasitic burdens, management burdens, maintenance burdens, etc.

Reduction of product energy, parasitic burdens, and control energy can sometimes lead to retrofit efforts requiring capital expenditures. A sound quantitative program, however, identifies for each expenditure the annual energy savings; then through rate calculations and forecasts of escalation, provides the anticipated annual cost reductions. The business executive or manager can then apply his own return-on-investment formula in selecting which options to elect.

The management and maintenance burdens are often found to be the greatest areas of energy waste. Examples include operating machinery when it is not needed, burning lights in unoccupied rooms, or using outdoor dampers that do not function properly. These losses can sometimes be significantly reduced by simply paying attention to the problem. Other problems may require capital investment in such things as a planned maintenance program, the installation of timing devices, or computerized management aids.

However, as with the retrofit expenditures, a good analysis will reveal the reduction in the energy consumption and resulting monetary benefit of each expenditure.

The professional's role

If the professional has done an effective job of educating himself and others, there is little doubt that responsible management agencies will seek him out. On a more active basis, however, he could simply ask his clients for a copy of their last year's energy bills. With some experience and judgment, and knowledge of the building and its energy system, the professional could tell whether there are any energy conservation opportunities and he is on his way.

The "sale" is the easiest part of the job. The hard part is developing a staff and technique capable of executing the comprehensive effort required in energy management. The staff requirement is considerably different than that needed for the production of system design documents. Energy management requires a combination of top talents in building materials and structures, systems analysis, management, maintenance, system testing, utility rates, report writing, and (minimal) drafting. Individuals of the staff or team may possess more than one of these varied skills, but they are all necessary. This is an entirely different inventory of skills than is usually found in the consulting engineer's office. Thus, the engineer must develop a staff or team capable of approaching the energy management market in a professional way. This may be economically painful initially, but the long-range benefits to both the financial stability of the firm, and the public, should prove well worth the investment.

As stated earlier, the professional firm makes or loses money when the contract is negotiated. Since every energy management program is unique, there is no common denominator to assess a fee basis. Furthermore, when such rigid fee structures are attempted, there has historically been a lack of clear definition by those proposing the fee basis as to what services are to be provided for the fee proposed. This leads inevitably to misunderstandings between the professional and the client.

The only way a professional can establish a fee for energy management consultation is to understand the extent of the involvement required, and determine a fixed or maximum fee on the basis of the anticipated time involved. As a limit, the energy cost savings potential must warrant the cost of the fee. If not, the scope of the project must be reduced to within the potentially available funds.

13

A chronology of building systems technology, 1929 through 1979

Our industry has made dramatic contributions to society; here's a look at what they have been and what they must be in the years ahead.

The early nineteenth century gave birth to an era known to every elementary school student, the industrial revolution. Along with the industrial revolution, the student readily identifies names such as Whitney, Watt, and Fulton. This was an era in which man learned to harness thermal energy and convert it to work, thereby amplifying his potential productivity many orders of magnitude. This vast increase in human productivity converted mankind's lot from a crude, almost primitive existence to the advanced society of comfort, accomplishment, and opportunity we know today. It is the *economic advances* born of the industrial revolution that are most commonly acknowledged.

Other names, not so readily recognized by the young student or the man on the street, are those of such men as Carnot, Rankine, Maxwell, and a host of other thermodynamicists of the nineteenth century. These names are generally recognized only by physical scientists and engineers. It was they who pondered the deep questions of the behavior of materials, fluids, and energy; identified these elements of behavior; and developed the science of thermodynamics. This science, in turn, formed the foundation of the industrial revolution, then went on to carry it into the twentieth century in high gear.

The primary goal of the vast majority of early machines (and consequently the direction of thermodynamic development) was to provide work or so-called shaft energy. This shaft energy was, and is, considered the highest form of energy; that is, the form with the highest "value." The next and related area of attention in thermodynamics was that concerned with machines that cause heat to flow from lower temperature sinks to those at higher temperatures—refrigeration machines. Then, to supplement the use of the thermodynamically produced shaft power, following discoveries by Oersted and Faraday in the early nineteenth century, electric energy was developed to transport the shaft level energy from place to place.

If we pick up the chronology of the development at the beginning of the twentieth century, we find that many exciting events were either occurring or were about to occur in the conversion of heat to work:

- Locomotives drawing long trains of people and cargo were crossing the continents, some containing their own power conversion plants and others utilizing power in the form of electricity.
- The factories of the world were humming with machines making everything from clothes

to wagon wheels to locomotives. Again, some of the factories had engines connected to line shafts while others used electricity provided from a nearby power plant.

- In the cities, there was no longer the need to travel by foot, horseback, or horsedrawn coach. Public transportation was available in the form of streetcars, powered by electricity or cables driven from central power plants.

- In western Europe and the United States, companies were being formed to manufacture machines that were actually intended to replace the horse—automobiles.

- Professor Langley's steam powered "aerodrome" had successfully flown for 1 minute and 49 seconds in 1896, and the race was on to develop a craft that would carry man through the air. (The Wright brothers accomplished the feat at Kitty Hawk, N.C., on December 17, 1903.)

- Candles and oil lanterns were being replaced throughout much of the Western world with gas lamps; electric arc lights were used in some cities to light the streets; and the successful development of the incandescent filament lamp was just around the corner.

- The process of generating steam from fuel, which had been developed for the then-commonly used steam power cycle systems, had been turned to another use, one not generally considered at the time to be a thermodynamic cycle. This use was as a heat transfer fluid, to heat buildings. The multiplicity of fireplaces and chimneys was giving way in many of the larger buildings to so-called central heating systems where the fuel (usually wood or coal) was converted to steam at a central point and the steam piped to the rooms and condensed in cast iron radiators to heat the space.

- Refrigeration plants were constructed in most larger population centers in order to make block ice from river water. The ice thus enabled people to preserve perishable foods— a concept that was about to change the eating habits of the world, significantly improving human diets, health, and quality of life.

From this beginning at the turn of the century, the continued industrial development to 1929 is generally recognized as being in transportation (the automobile and the airplane) as well as in industrial processes, which were turning to mass production. Public attention to the depression, the related economic hardships of the 1930s, the war in the 1940s, and the cold war in the 1950s drew attention from the continued development of the industrial revolution. An aspect of this period that generally comes to mind, however, is the conversion of the industrial war machine of World War II to civilian purposes.

Some of the evident highlights of this conversion took a notable turn from the trend of the prewar era. In rail travel, the 1940s and 1950s saw the steam locomotive replaced by the diesel-electric and also saw the rail passenger business virtually disappear. Very little happened with postwar development of the private automobile, except for slow modernization with increases in size, increases in horsepower, and increases in fuel consumption. (It must be recognized, however, that improved manufacturing processes brought on the era of two-car families.) The most recognized change was perhaps in aircraft development and air travel. The propeller plane was replaced almost totally by jet powered aircraft, until today we have gigantic jumbo jets that hold hundreds of passengers and fly just under Mach I as well as the European Concord that crosses the Atlantic at supersonic speeds.

But behind the romance . . .

These are the "romantic" remembrances of the continued impetus of the nineteenth century industrial revolution in the mid-twentieth century. But there was another branch of this development that had an even greater impact on most of our lives and provided the catalyst for further improvements in productivity and science. (Keep in mind that increases in productivity per capita are the *only* way mankind can improve its lot!) This branch of development was in the *environmental sciences*. This area of technology probably had the least time in development, the least research or development supported by public monies, and yet the greatest impact on our lives and life styles of any branch of science or engineering during the twentieth century.

Central heating systems advanced slowly from the nineteenth century through the World War II era, picking up other central heating concepts along the way such as gravity hot water heating and warm air. In some cases, in large buildings, systems were combined, particularly to address the need for ventilation. But by today's standards, the systems of a mere 35 years ago were primitive to say the least. Around 1911, Willis Haviland Carrier, an engineer whose name and contributions are little known to those outside the industry, initiated the twentieth century advances in the science of conditioning the air (prior work had been documented as early as 1837, but little use for the mathematical relationships was recognized).

Carrier's work grew out of the cotton mills and other limited industrial applications prior to World War I and, supplemented with the thermodynamic refrigeration cycles of the previous century, started creeping into the eyes of the public during the post depression era of mid and late 1930s. In those years, theaters and other places of public assembly such as large hotel ballrooms and meeting rooms were seen to be advertising "20 degrees cooler inside." The prior option was that such spaces simply were not used during hot weather. Parallel activities during the same era found the household ice box being replaced by a refrigeration plant, not on the river but right in the kitchen!, and the centrifugal pump applied to the gravity hot water heating system to make it more forgiving of pipe sizing, balance, and response problems.

These developments ceased during the war years, as did those in most nondefense-oriented fields of endeavor. The post war years saw environmental systems develop like a new industrial revolution, with nearly as significant a social and economic impact as the nineteenth century industrial revolution.

Some engineering and business talents were directly convertible from military to civilian activities and were channeled into such fields as transportation and communications. Others were seeking a new home, and with no involvement or direction from the federal government, United States society found a slot for technologists in the little recognized area of environmental control. Admittedly, in the manufacture of explosives during the war, the recognition of the integration of space cooling and heating had developed. Thus was born the concept of space air conditioning as we know it now.

How we contributed

It is almost awe inspiring to reflect upon the impact the building environmental industry has had on United States economic growth following World War II and on other advances in science and industry. And it is interesting to note that, scientifically, there have been no so-called breakthroughs; the entire process has been one of ingenuity in application engineering and product development.

The United States economy has been basically inflationary since 1946. Energy sources developed to serve the United States defense machine during the war were made available to the civilian market at a very modest price. This availability tended to exert market pressures on those areas of technology that could benefit the economic system and the social welfare while retaining the needed growth in the energy supply industry. The response was that a total change occurred in the methods of constructing indoor space.

The architect was released from the prior comfort-related constraints of mass control, ceiling heights, window designs, and the very shape and siting of the building. He was totally free to shape and arrange the space to suit the functional needs of the occupants. The availability of low-cost energy and the technological advances in the engineering disciplines that comprise the building environmental systems industry combined to provide the necessary environmental control irrespective of the envelope shape or material. Not only did this technology enable the industry to create more comfortable spaces regardless of weather or building location, but the improved space quality was also provided at a lower real cost per usable unit of area.

The effect this had on the economy is analogous to increased productivity: we were getting more for less. Thus, the building industry, conceptually, was a high-productivity industry that contributed to a growing economy rather than to inflation!

The far-reaching effects of the revolution in building designs and technology are difficult to identify and summarize, but the following is an attempt to recognize a few of the more significant contributions:

• On July 20, 1969, the United States Spaceship Apollo 11 landed on the moon, and astronauts Neil Armstrong and Ed Aldrin walked on the moon's surface. All mankind recognizes this as a magnificent accomplishment of the aerospace industry, but a key component of that industry is the building environmental component. Without specific and substantial contributions from the environmental sciences, the feat would not have been accomplished. One such contribution was the extensive research on human comfort and human heat dissipation (much of it sponsored by private industry through ASHRAE). Another was the ability to provide essentially sterile atmospheres with extremely accurate temperature and humidity control, which were necessary for the manufacture of sensitive electronic components and for subsequent assembly of these components.

Health care advanced

• The field of health care has certainly been advanced by developments in heating, ventilating, and air conditioning. Modern surgical suites are dependent on high-efficiency filtration and pressurization techniques that remove airborne dirt and bacteria from entering air and prevent infiltration of contaminated air from adjacent areas. Precise humidity control also provides an effective means of maintaining cleanliness since the growth rate of microorganisms is minimal at approximately 50 percent RH. Also, stable temperatures and humidities are important for patients in burn treatment centers and intensive care units as well as those weakened following surgery or other trauma.

On a less sophisticated level, modern air conditioning systems have provided relief to millions of sufferers of allergies and respiratory problems. And it would be virtually impossible to estimate the number of cases of heat stress and other medical problems that have been prevented by air conditioning and ventilating systems in our homes and workplaces.

Enter computer technology

• Computer technology touches on virtually every phase of our lives. Computers read the prices of the goods at the grocery store and department store; they calculate the monthly bills and invoices for virtually all industries from the family doctor to the credit card agency; they schedule industrial production and control inventory; they calculate our paychecks and bank balances; they categorize real estate listings. In virtually all occupations, we find ourselves interfacing (if not competing) with some form of computer technology.

But to do all this "thinking," the computer appears to be a second law machine (much like man) and must therefore dissipate or reject heat. As such, the computer is quite sensitive to its environmental conditions of temperature and humidity. It could never have been developed in the manner and within the economic structure that it was were we not able to provide the computer with an environment of closely controlled limits of air temperature, humidity, cleanliness, and distribution.

• The economic effect of computer technology and use development, properly applied, was and is to increase the productivity of man. But consider a more direct relationship between environment and man's productivity. After a good night's sleep in the controlled temperature of a modern home, a person is as physically and psychologically fresh at the start of a workday following a 90 F humid night in August as he is following a "perfect" 60 F evening in spring or fall. The office worker can cope with his production requirements equally well year round. *It is not simply a matter of comfort.* Consider the problem of an engineering or architectural office trying to produce tracings for blueprints during warm, humid weather in the years preceding the quality of air conditioning known today. Not only did extreme discomfort tend to reduce productivity, but the mundane challenge of protecting the tracings from perspiration smears consumed as much time as putting the lines on the paper. If a truly scientifically accurate study addressing all of the relevant considerations were undertaken, it would not be surprising to find that summer productivity (in the moderate four-season zones) has increased at least 100 percent as a result of air conditioning technology.

Air conditioning opens sun belt

• In some areas of the United States, the period of thermal discomfort for indoor activity was not a simple seasonal problem but rather a continual condition. These areas included much of the Southeast, South, and Southwest. Because of the indoor discomfort due to high temperatures, these areas were predominantly rural in nature, and the major industries (except for mining and petroleum drilling) were farming and ranching. The first decades of the past 50 years saw an overabundance of farm products, which, even with governmental support, pointed to a certain economic decline of those areas.

Just as the decline was imminent, the new building technology freed the sun belt from decline and opened up a future of dramatic growth. Not only did it provide opportunities to increase productivity significantly through both residential and commercial air conditioning, but it also provided the vehicle for a massive movement of people and industrial activity from the congested areas in more northern sectors of the nation.

Auto cooling aids market

• Air conditioning and refrigeration technology has contributed more to the expansion of the automobile market than any other single contributor. Early efforts, like those in most commercial developments were add-on cooling units, which started to impact the market in the early 1950s. Then the auto manufacturers saw a public demand and began to integrate cooling systems into auto design to provide year-round indoor comfort control in the family car. An initial observation is that this "accessory" simply makes the ride in the car a bit more comfortable during hot weather. The fact is that a few days to several weeks of driving on a family vacation in hot weather without a cooling system was an almost unbearable experience that relatively few people took to. It was the addition of cooling that sent a vast number of American families across the nation's highways during the school vacation months of June, July, and August.

A natural next step in the evolution was the advent of the recreational vehicle, a home wherever one chooses to go. Without current air conditioning technology, the RV as it is known today would never have existed.

Another industry that was virtually built upon the air conditioned car is the modern motel industry. Increased numbers of travelers, needless to say, required places of rest and refreshment, and to provide for this need, a whole new industry of modern motels grew up along the nation's interstate highways, which are themselves at least partially attributable to the air conditioned car.

GNP per capita climbs

• The best measure of standard of living is thought by some to be the adjusted GNP per capita. Another measure of what might be more correctly called quality of living is the percentage of personal income spent on leisure activities. This ratio has continually climbed during these past 50 years with few temporary declines. Consider the impact that advances in HVAC technology have had on leisure activities, starting with bowling. Admittedly, the bowling pinsetting machine was a catalyst for the growth of the industry in the 1950s. But without the year-round revenues from use of the machines during all outdoor temperatures, it is doubtful that the industry could have supported the cost of the pinsetters. In other areas of leisure time activities, one of the major recreation growth industries in the past decade has been the racquet sports of tennis, racquetball, and squash, all played now indoors in controlled environments. Hockey, both professional and amateur, can now be played at any time of the year in Houston, Texas as well as Ottawa, Ontario because of the advanced technologies of refrigeration and air conditioning. Even the massive spectator sports of football, baseball, and soccer are played in indoor "stadiums" in some places where it tends to be too warm for outdoor baseball in summer or too cold for outdoor football in winter.

A plan has even been developed to provide indoor facilities for skiing in areas of the country that do not have the needed mountains and temperatures.

These have been but a few brief examples of the impact that the air conditioning and refrigeration industry has had on society over the past 50 years. A key element in the dissertation has been that the technological growth was a response to needs of the United States socioeconomic system following World War II. As previously stated, the needs were:

1) A job market was required for scientists, engineers, and technologists following the war effort.

2) The energy industry, which had geared up for massive productivity during the war, was pressuring society to create a market for their goods so that their industries could continue on the necessary growth curves.

The next 50 years

As we in the industries of building environment and refrigeration embark on the next 50 years, we might look proudly on our contributions over the past half century, but might then reflect on whether there is to be a change in the previously set course.

It was mentioned that the accomplishments achieved, initiated to satisfy society's needs as stated above, came about through free market pressures. There was no master plan, so to speak, developed by the government or our political leadership. This observation is most significant as we glance forward to the coming half century.

Many of today's practitioners in the industry are those who have had the opportunity to participate in the industry's growth in the past

50 years. It now appears that the ground rules or society's needs have changed and that we will have the opportunity to lead in the solution of a new problem. It is doubtful whether *any* given sector of technology has ever had such an opportunity to assist the socioeconomic system on so significant an undertaking before; *and in this industry, we are about to be given the opportunity for a second time within five decades.*

To address the needs of our emphasis and directed energies for the coming half century, compare them to the above-stated needs of society following World War II. Coincidentally, the first need, that of retraining, has run a complete cycle and exists once again. This time, the federal government, bent on solving all social needs, is attempting (with little fruition) to handle the retraining through government contracts and by directing thousands of former aerospace and AEC scientists and engineers into government sponsored programs concerned with building environments, most of which have had, or show promise to have, little effect on the industry or society. More quietly, there has been a personnel shift of aerospace scientific and engineering talent into the free enterprise market of the environmental and refrigeration industries. This is a time when the industry needs new talent at a greater rate than our educational system can provide, for a reason that may not be immediately evident and that leads to the second need of today.

In the late 1940s, the energy supply and production industry had a need to grow—a need that was real—and the resulting growth catalyzed an unprecedented period of economic growth. This situation has changed markedly.

Growth problems recognized

There were problems with this growth that were fairly well recognized by all knowledgeable practitioners in energy systems. The vast majority of the citizenry and, unfortunately, many of our political leaders, however, were not cognizant of them. A most timely observation came in a paper presented by G. W. Gleeson, dean of engineering at Oregon State

College, at the Semiannual Meeting of ASHVE in July 1951.* His paper, entitled "Energy—Choose It Wisely Today for Safety Tomorrow," presented an in-depth study of the then-current use patterns of energy resources (coal, oil, oil shale, natural gas, nuclear, hydro, vegetation, solar, wind, geothermal, tidal, tropical waters, and heat pump potential) and compared them to the resource availability. The paper illustrated clearly that continued dependence on United States resources in energy (particularly oil) would lead to a short-lived growth economy. A similar message related strictly to petroleum was delivered by Dr. M. King Hubbert in his address to the American Petroleum Institute on March 7, 1956.

The oil industry responded to the situation by seeking other sources and commenced rapid development of oil resources in other parts of the free world, including South America, the Middle East, and other relatively undeveloped areas. Thus, with the full knowl-

*Gleeson, G. W., "Energy—Choose It Wisely Today for Safety Tomorrow," *Transactions, The American Society of Heating and Ventilating Engineers*, Vol. 57, 1951, pp. 523–540.

edge and blessing of the United States government, we continued our growth economy on borrowed time while becoming the world's largest oil importer, until today we are dependent on foreign petroleum products for the survival of our economy.

Our leadership, instead of recognizing the need for redirecting our efforts in technology, urged us to go on spending with programs such as "atoms for peace," with one of the slogans informing us that with this new energy technology, electricity would be so inexpensive we would not need meters!

Now we must conserve!

We must now as an industry redirect our efforts toward the goal of preserving for our society all of those benefits that we have provided, such as those enumerated above, while reducing the rate of consumption of energy resources. If this can be accomplished, it will be an economic achievement of the first order.

Figure 13-1 is a graph of GNP per capita versus energy consumption per capita for the period of 1909 through 1973. An observation

Per capita
annual energy
consumption
10^6 Btu

Per capita
GNP $S10^3$
(1958)

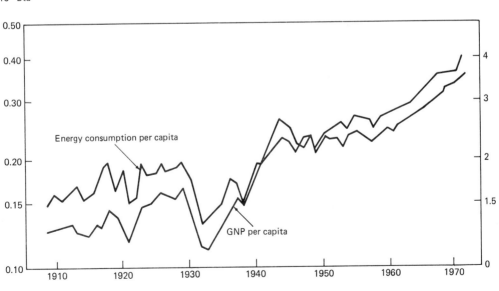

Source: Institute of Gas Technology

Fig. 13-1 Trends of per capita GNP and annual per capita energy consumption in the U.S. 1909–1973

of the near congruence of the two curves leads to the inevitable conclusion that unless non-depleting sources can be found to replace our present depleting sources, *in equivalent quantities*, we have but one choice in charting our future course of action: that is to solve the problem of direct interdependence between energy consumption and GNP.

This can be done through more judicious use of energy. We can design and build homes and buildings in such a way as to provide high levels of comfort while consuming less energy.

We can market frozen and perishable foods effectively while consuming less energy to do so. We can run our gigantic industrial system on a fraction of the resource energy we are consuming.

There will be relocation of people from one industry to another, just as personnel shifted from the railroads to the airlines in the late 1940s and 1950s; the appearance of buildings will change, just as it has over the past 50 years. It will be an exciting time. Our successes are behind us, and the challenge is before us!

14

Can we afford engineering?

Other chapters discuss energy economics, design parameters, and maintenance. In proper perspective, energy economics and maintenance are two design parameters relating to building systems. Still others address more parameters that should be considered and some of the manners in which they should be viewed. Few will disagree with these suggestions. However, it is anticipated that most system designers will pose the question: In-depth consideration of all these parameters will produce a better system design, but, realistically, does the present system of financial compensation for consulting engineering services in building systems allow for all these considerations?

To properly address this topic, let us review the events leading to the present situation and some of the marketplace adjustments that have evolved recently as an inevitable response.

Before World War II, most commercial and institutional buildings had only heating and ventilating systems. Heat was provided primarily by steam or two-pipe hot water, standing radiation; ventilation was by either gravity or fan motivated exhaust systems (in rare instances, heated makeup air was provided).

These systems were generally designed by manufacturers of steam specialty devices or their sales representatives. In an effort to aid the owner in selecting a system to best suit his particular application, the consulting mechanical engineering profession emerged. Many early consulting engineers came from manufacturers' design departments.

During this time, fees were established as a percentage of the architectural fee, which was based on a percentage of the construction cost, and they were equitable compensation for the work involved. Even after the consulting engineer assumed the added responsibilities of laying out the piping-radiator-boiler system and specifying the components and materials, this fee structure proved adequate for the time and skills involved.

After the war, a fantastic technological evolution in building environmental systems occurred. This technological boom through the 1950s was surpassed by no other nonsubsidized effort in history. This technology made it possible for architects and the building products industry to have almost complete freedom of design in building shapes, space functions, and materials. It also resulted in extreme complexity in, and dependence on, the mechanical systems.

Fee structure inadequate

In spite of these extensive changes and the resulting increased demand on the skills and time of the mechanical systems designer, the concept of the design fee as a relatively small percentage of the system construction cost has remained in most contracts. As a result, for financial survival, most design firms had to hold services to a level dictated by the available fees. Thus, standardized office practices that devoted the most time to layout rather than costly in-depth analysis became the rule rather than the exception.

Understandably, this has caused less than desirable results in satisfying many of the design parameters. Recognizing this, the building industry has sought other means of obtaining more thorough engineering considerations under such labels as turnkey, systems, package, etc. All of these are simply methods of providing the desired and needed engineering services as a component cost of the construction contract.

Fee specialist consultant

Another method that evolved from this is the specialist consultant, such as the value engineer (first cost specialist), energy specialist, corrosion engineer, water treatment consultant, acoustical consultant, etc. Generally, these specialists are employed by the owner outside of the architectural contract.

It cannot be denied that either approach provides; if properly executed and applied, a means of funding more adequate engineering fees. However, it should be recognized that the competent, independent professional with no profit motivation in a product, and with an equal interest in the balanced integration of *all* the design parameters, can best serve the owner's interests. One problem: How will he be compensated for his efforts?

Complexity increases cost

In summary, the cost of engineering building environmental systems has increased enormously due to increased complexity. Consider the cost burdens of computerized load analyses, life cycle cost studies, maintenance requirement analyses, energy studies, distribution systems analyses, and energy source analyses—all part of the analytical requirements before a line is drawn on a tracing. Consider further, that these aspects of the design process require the skills of highly educated and experienced professionals.

Add to all of this the inverse motivation built into the percentage fee contracts—increased engineering hours devoted to cost reduction, resulting in a reduction of compensation—and the question must be asked: Who is the beneficiary of this arrangement?

This essay does not propose answers, only the question. Properly engineered building systems require monetary investments in engineering. It is time the industry faced this fact and sought an answer!

SECTION II

Concepts of energy

As this book is being assembled, the world can be considered as being in the throes of an energy revolution. The industrial revolution which transpired from the early part of the nineteenth century through the middle of the twentieth century is described by Webster as "The change following and resulting from the introduction of power-driven machinery to replace hand labor, ..." The substitution of such machinery for hand labor has had the direct effect of amplifying by many orders of magnitude man's productivity, and has been the absolute foundation of the world's economic development for these past two centuries. Like population patterns left unchecked, the progression of such "power-driven machinery" follows a geometric pattern leading to a point which in population growth is called an explosion. In industrial development, the progression curve reached this point in the mid-twentieth century, and the pressure points manifested themselves in unexpected ways. The early signs were those of atmospheric and water pollution. In earlier times the problems of pollutants were much more localized, and were solved by physical separation between places of habitat and points of pollution, or by changing from the polluting process to a "cleaner" one. The energy conversion processes are now so extensive and universal, that the pollution problem is virtually unsolvable. The financial burden associated with the solutions that have been attempted has the effect of reducing productivity thus damaging the very economy that is being supported by the energy conversion.

The other pressure point which has been recognized by some scientists and business leaders for some time, but only became evident to the major sectors of the populace during the decade of the 1970s is that of energy resource limitations. The geometric progression of the power-driven machinery put enormous strains upon the supplies of energy that fueled that machinery—fossil fuels. The process of controlled combustion—fire—has been used by mankind since his earliest existence on earth, and it is still the process that we use to motivate our industrial revolution. Efforts at replacing the combustion process with more sophisticated (nuclear) or continually replenishing (solar) processes are providing a small portion of the needs, but in comparison to the whole these are quite limited.

Thus, it appears that we are approaching the point of diminishing returns in energy. This is the point where the human efforts and resource efforts required to extract, transport, refine, and convert the energy requires greater productivity than the productivity increase provided by the last increment. If this is truly the case, the productivity benefits of the industrial revolution will cease. It is for this reason that the energy dilemma of the 1970s is referred to in the opening sentence as the energy revolution.

The chapters in this section are directed not at specific aspects of the energy situation or any proposed solutions, but simply are intended to provide some food for thought— perhaps giving the reader a new or different viewpoint. The chapter, "A Primer on Energy" is an elementary discussion on energy and power which was initially published for a nontechnical readership. It is sometimes beneficial, though, for all of us to go back to the beginning to reestablish our foundation for such complex topics as energy. The chapters on "A Definition of Energy," "Energy Is a Unique Commodity," "Energy Transportation," and "Infinite Source," are intended to provide some food for thought.

The closing chapter, "An Energy Resource Standard," is worthy of a brief introductory discussion. At the time this work was originally published in May 1977, The American Society of Heating Refrigerating and Air-Conditioning Engineers (ASHRAE) had been working for about three years trying to achieve consensus on a consensus standard (see Section V) for the design of energy-efficient building systems. The chore appeared quite straightforward at the outset. However, as the authors attempted to address the fact that building systems commonly can and do receive their energy from more than a single source and in more than a single form, it was necessary to seek a common denominator. As discussed in that chapter, this problem had not at the time of publication been resolved. This is an extremely complex problem to which many people, institutions, and agencies have been addressing themselves since the issue first came to public attention through ASHRAE's efforts. At the time of this writing, an acceptable solution has still eluded us.

15

A primer on energy

Generally speaking, all energy on earth comes from the sun. Energy exists in different forms, such as chemical, nuclear, sound, light, mechanical, heat, etc. Energy can be converted from one form to another, and all energy eventually converts into heat when being expended.

What energy is

Energy can be defined as the ability to do work, and work is defined as expending of a force through a distance, for example, lifting an object to a higher elevation. If a box weighing 10 lb is lifted from the floor onto a table which is 3 ft above the floor, the work required is 10 lb times 3 ft or 30 ft-lb. To do this work, it will theoretically require 30 ft-lb of energy.

Conversion of energy

A law of physics known as the law of conservation of energy states that energy can neither be created nor destroyed. This law is very important in understanding how energy is used and how it can be conserved. The 30 ft-lb of energy that was required to lift the box onto the table is not used up and lost; rather it is stored at the box as "potential energy." If at a later time, the box were lowered from the table back to the floor, the 30 ft-lb of stored potential energy would be available to do some other work, such as lifting an object to another height assuming box and object are connected to a system of cords and pulleys. An example of such a device is the double hung window with counterweights in the window jamb. When the upper window is pulled down, counterweights rise within the jamb; force expended by the person to pull down the window is stored as potential energy onto the counterweights. When the window is again raised, energy stored in the counterweights provides most of the required work to lift the sash.

Aside from the direct work energy, there are other forms of energy. Most people have heard of the Boy Scout's technique of starting a fire by rubbing two sticks together. This technique uses different forms of energy. It takes work (and thus energy) to rub the sticks together because of a resistance to the "rubbing" called friction. That is, the sticks are not completely free to move against one another when they are pressed together; it takes energy or work to make them slide. But this energy is not stored as was the energy used to lift the box. What happens is that the sticks get hot, and work energy is thus converted to heat energy. Heat energy is measured not in foot-pounds, but in British thermal units called Btu. It requires 778 ft-lb of work to convert to 1 Btu of heat energy. In this manner, the work energy of rubbing sticks is converted to heat energy. When the heat energy accumulated becomes hot enough, it could ignite some combustible materials, such as wood chips, and the fire is started by a process of combustion. In the combustion process, chemical energy which is stored in the wood chips combines with oxygen in the air and releases thermal or heat energy. The chemical energy stored in the wood is useless until it is released through this combustion process into the form of heat. Once the heat is released, energy can be utilized for space and water heating, cooking, and other processes.

Gasoline used in an automobile engine ignited by spark plugs in a mixture with air (containing 21 percent oxygen) converts chemical energy to heat and work energy providing the mechanical force to move the car.

Table 15-1.   Energy value of common fuels.

Fuel	Approximate high heat value
Natural gas	1,000 Btu/cu ft
Propane	93,000 Btu/gal
Oil	140,000 But/gal
Coal	10,000 Btu/lb
Wood (12 percent moisture)	8,000 Btu/lb

Fuels classified

The materials used to release stored chemical energy by combustion are called fuels; most common fuels contain as basic ingredients carbon and hydrogen, or hydrocarbons. There are two common classifications of fuels—bio-fuels and fossil fuels.

a) Bio-fuels are products of growing matter such as trees and plants. This matter receives energy from the sun which is combined with other chemicals of the earth as they grow over a relatively short time span. At some stage of growth, these bio-fuels can be harvested and used for fuel with replacements planted for a later harvest.

b) Fossil fuels are generally considered as nonreplenishable and therefore depleting. Fossil fuels also derive their energy from the sun, but it took millions of years for them to form to their present state. Common fossil fuels are natural gas, petroleum, and coal.

The amount of energy in the form of heat that can be obtained by the burning of fuel is called the high heat value of fuel; some of the most common fuels and their approximate high heat values are shown in Table 15.1.

Electrical energy

Another familiar form of energy is electrical energy. Electrical energy provides a very useful method or means of moving energy from one place to another, or of transferring from one form to another. Electrical energy can be transformed from chemical energy (fuel cell and batteries), nuclear energy, heat energy (thermoelectric and thermionic), magnetohydrodynamic mechanical energy (hydro), and thermal-mechanical energy. Over 90 percent of the electrical energy generation in this country is by thermal-mechanical method where electrical energy is produced by driving a rotation electrical generator with a steam turbine. The steam turbine derives its energy from expanding steam produced by burning fossil fuel. In this energy conversion process, chemical energy is converted into heat energy, to mechanical energy, and then finally to electrical energy. Electrical energy is then sent through transmission lines to distant locations where it is converted to other useful forms, such as running a fan (to mechanical energy), heating a space (to heat energy), and lighting a task (to lighting energy).

Electrical energy is usually measured in kilowatt-hours (kW-hr or KWH) where 1 KWH is equivalent to 3413 Btu. Another energy unit commonly used is "horsepower-hour" (hp-hr) where 1 hp-hr is the equivalent of 2545 Btu.

Energy units

Common forms of energy used or found in building systems and their units of measurement are shown in Table 15-2.

The British thermal unit (Btu) is the most commonly used unit for energy measurement or for a common base of energy conversion; however, it is a very small unit (when compared to the amount usually required in building systems). In building energy evaluation, one must constantly deal with astronomically large numbers. Frequently, hundred-thousand Btu (therm), million Btu (mega-Btu), billion Btu (giga-Btu), or quadrillion-Btu (quad-Btu) are used for expressing annual energy consumption.

Power and energy

Power has often been confused with energy. In trying to understand energy conservation and energy management, it is very important to understand the difference between these two terms.

Power is the rate of consumption or conversion of energy; that is, power is an expression of how long or how quickly a given amount of energy is consumed or converted. An example: if 100 hp-hr of energy is used in 1

Table 15.2. Energy equivalent.

Form	Unit of measurement	Btu equivalent
Heat	British thermal unit (Btu)	—
Electrical	Kilowatt-hour (kW-hr)	3413
Mechanical	Horsepower-hour (hp-hr)	2545
Chemical (fuel)	Pounds, gallons, cubic feet	See Table 15-3

hr, the power required would be 100 hp-hr divided by 1 hr, or 100 hp. Thus a motor or engine to provide energy at this rate would be a 100 hp motor; however, if the same amount of energy were used over a 10 hr period, it would require only a 10 hp motor. Expressed in a mathematical equation, the relation between power and energy is:

$$power = \frac{energy}{time}.$$

Power units defined

If each energy term discussed previously is divided by time, then the power unit for heat, mechanical, and electrical energy will be Btu/hr, hp, and kW, respectively.

If a building is being cooled, it is necessary to remove heat (energy) from the space. This cooling procedure is more commonly called "air conditioning." The rate at which this heat is removed can be expressed as Btu/hr cooling, or tons of refrigeration. Common power units and their equivalent values in Btu/hr are shown in Table 15.4.

The effect of power on costs

Consider the case of the 100 hp-hr of energy consumed in the earlier example. If it is feasible to use a 10 hp motor for 10 hr per day, it would certainly be better than using a 100 hp motor for only 1 hr per day because the investment cost of a 100 hp motor is much higher than the 10 hp motor even though they both consume the same amount of energy (100 hp-hr) per day. It can therefore be said that power is related to investment cost, and energy is related to operating cost.

Electrical usage for a building is generally charged by the amount of energy (kW-hr) consumed per month. However for large users, the utility company may also base its charge on power (kW) demand which is normally determined by the maximum demand during any 15 min interval in a month or year. In this case, power demand is also related to operating cost for the user. The

Table 15.3. Energy unit (EU) value of common sources.

Type of energy	Unit of measure	Approximate value*	
		English unit (Btu)	Metric unit (kcal)
Electricity	Kilowatt-hour (kW-hr)	3,413	860
Gasoline	Gallon (gal)	128,000	32,000
Fuel oil (no. 2)	Gallon (gal)	140,000	35,000
Residual oil (no. 6)	Gallon (gal)	147,000	37,000
Natural gas	Cubic feet (cu ft)	1,000	250
	Therms (tm)	100,000	25,000
LP gas (butane)	Gallon (gal)	102,600	25,600
LP gas (propane)	Gallon (gal)	93,000	23,300
Steam (at 14.7 psia)	Pound (lb)	1,150	290
Steam (at 200 psia)	Pound (lb)	1,200	300
Coal (average)	Pound (lb)	10,000	2,500

*Btu—British thermal unit; kcal—kilocalorie. Energy values shown in the table may vary with source.

Table 15-4. Value of units.

Form	Unit of measurement	Btu/hr equivalent
Heat	Btu per hr	—
Electrical	Kilowatt	3,413
Mechanical	Horsepower (hp)	2,545
Cooling	Tons of refrigeration	12,000
Chemical (fuel)	Btu per hr	—

utility company, however, must charge for the demand (power) to offset their investment cost in the generating plant and distribution network.

Energy required in buildings

Energy is generally used in buildings to perform functions of heating, lighting, mechanical drives, cooling, and special applications. The energy is available to the building in limited forms, such as electricity, fossil fuels, and solar energy, and these energy forms must be converted within the building to serve the end use of the various functions. A loss of energy is associated with any conversion process. In energy conservation efforts, there are two avenues of approach—reducing the requirement of the end use, and/or reducing conversion losses. The latter is an unfortunate situation inherent in most conversion processes. For example, the furnace which heats the building produces unusable and toxic flue gas which must be vented to the outside of the building, and thus part of the energy is lost. (The "lost" energy is not destroyed, it simply ends up as heat energy.)

Conservation of depletable energy

The ultimate source of energy, as stated earlier, is the sun. It is plentiful and nondepletable for million of years; however, from the perspective of current technology and social systems, the vast majority of energy resource is still fossil fuel, which is a depletable commodity. Even though electricity can be converted within buildings to other forms of end use with relatively low losses, it has already undergone a conversion process in which approximately two-thirds of the fossil fuel energy was lost, unless electricity is generated by hydroelectric plants where energy is converted by controlled water fall (potential energy).

Solar energy, the only continually nondepleting source available, is not easily converted to the form needed for the end use, and the investment cost of solar energy conversion systems is quite high compared to more conventional sources. In achieving the goal of energy conservation, efforts must be directed toward minimizing consumption of depletable energy resources.

16
A definition of energy

The concept of a subscience of energy economics is presented in some detail in Chapter 21. Since engineering is an applied science founded to a major degree upon the basic science of physics, engineers tend to think as applied physicists. A natural consequence is the tendency to seek exact definitions when addressing engineering problems.

This chapter will consider the seemingly parochial problem of defining "energy." Doubtless, there are few engineers who do not know what energy is; just as there are few Americans of any walk of life who must even stop to think of what energy is as they listen to energy-related news reports or read endless references to energy shortages, crises, and related problems in the daily newspapers.

Most engineers will probably think back to the fundamental definition in introductory physics: "energy is work or the capacity to do work."

Energy is capacity to do work

Those with a less technical background will form mental images of energy, which take such diverse forms as gasoline for the family car, warm homes in winter, cool homes in summer, electric bills, gas bills, active children in constant motion, etc.

Whether one relates to the fundamental physics definition or the less structured concept, our society has managed to develop a frightening dependence upon the energy potential stored in the earth's nonreplenishable resources without the benefit of exactly defining the source of our dependence. Partially, as a result of this, we find ourselves in the midst of an energy consciousness that has been called everything from a crisis to a mere situation and which has generated the expending of literally tens of millions of man-hours

and billions of dollars in discussions, studies, research projects, etc., resulting to date in numerous documents drafted and published ranging from policies, standards, and guidelines to federal, state, and municipal laws.

Other definitions of energy

After reading a number of these documents, it becomes evident that if the classical definition were applied to the word energy as used therein in the legal context, it would likely be thrown out in the lowest court. Let us then consider a definition of energy as it might relate to the so-called popular movement of energy conservation or the science of energy economics. A brief scattering of some readily available sources reveals the following definitions (or lack thereof) of the word, energy.

- *Webster's Dictionary:* "capacity for performing work."
- High school physics text: "the capacity to do work."
- Engineering thermodynamics text (Stoever): ". . . the something that is transferred to or from a system: a) when work is done on or by the system, and b) when heat is added to or removed from the system."
- Introductory economics text (Samuelson): no definition.
- State of Minnesota Legislation HF No. 2675, an act relating to energy: no definition.

From this brief sampling, it is seen that little has been done to define energy beyond the basic physics definition. Yet, when we start discussing energy in terms of energy economics or energy conservation, most will agree that we are addressing the conservation of tangible things—not simply forces, distances, and relative temperature levels.

Stored energy must be converted

Considering the concerns of energy resources available to mankind, starting at the practical source, all energy on earth emanates (or emanated) from the sun in the forms of nuclear and thermal energy transmitted in mass migration with the formation of the solar system, or on an ongoing basis by radiation waves. Following many centuries since its formation, the earth has stored a given amount of the original mass migration energy in relatively fixed or limited quantities. This stored energy is available for conversion to useful forms, generally through chemical conversion processes, then through the first law processes (building heating and industrial thermal needs), or second law processes (transportation, electrical power generation, refrigeration, shaft power). It is these sources that are the rightful target of so-called energy conservation efforts. Few would deny that efforts in energy conservation do not relate to reducing the energy available from the sun, the tides, or the human energies of mankind.

Conservation reduces use rate

Thus, efforts at energy conservation are actually efforts at reducing the *rate of use* of the limited or finite-energy-producing potential stored within the earth. All attempts at the control or reduction of energy usage or consumption should therefore address a definition of energy relating to this consumption or depletion of the nonreplenishable resources of the earth regardless of the quantity that is stored. Any finite quantity will eventually be depleted.

Consider then the following definition of energy as it would relate to energy conservation efforts, energy economics, or energy policies: *"Energy is the potential for providing useful work or heat stored in the finite resources of the earth."*

Although it may appear that the various available forms can be totalized by reducing to a common denominator, such as the potential heat content (say, Btu), the study of energy economics must take into account the available quantity of each form, and the convertible use for the given forms by the current state of the art technology.

The resources, within the current context of known technology include three general forms (which could be further subdivided): fossil fuels, nuclear, and geothermal. Other chapters address the issue of the value of the various forms in which the resources are found.

17

Energy is a unique commodity

Such topics as energy economics, energy management, and energy conservation have been discussed broadly. Lately, it has become increasingly apparent that another fundamental of energy should be considered—the uniqueness of energy as a market commodity. In Chapter 21, the definition of economics is quoted from Webster: "a social science concerned chiefly with description and analysis of the production, distribution, and consumption of goods and services." In light of this definition, it should be noted that energy resources, like other natural resources that support the monetary system, can be and have been considered as goods.

As goods, or a commodity of commerce, energy resources must be considered quite apart from all other commodities, and entirely different rules of monetary economics must be applied if social systems as we know them are to survive. The more advanced socioeconomic systems as they exist today generally consider energy resources in the same category in which other commodities of commerce are considered. Yet energy resources are the single commercial commodity upon which the economic systems are structured.

Man in earlier times recognized the value of energy resources in a way in which modern educated societies do not. Earlier societies, both primitive and advanced, in many cases considered energy resources as a god—the sun god, the god of fire, etc. Today, countless polls taken in the United States have revealed that the average citizen does not think there is an energy problem, only a ploy on the part of business institutions to increase profits. Many believe, in the search for a solution, that the *other guy* should be doing something to conserve. We would do well to consider the wisdom of those earlier societies.

Energy is fundamentally matter itself, an observation that Einstein quantified in his energy-matter equation. *No* other resource or commodity approaches this fundamental position.

Energy resources compared

In comparing energy resources with other resources extracted from the earth, one can use the metal resources as an example: iron, copper, aluminum. Several differences immediately become apparent. Although we have based immense sectors of our socioeconomic structure on these resources, none of them could have been processed into a useful form without the assistance of energy resources. Although we have had each of them available and have utilized each of them for maximum economic benefit, the vast majority of products and goods produced that use them could have been developed eventually by using another metal. Furthermore, in the majority of product applications, the resource material, properly protected, did not deplete to an unusable form, i.e., the automobile made of steel could be recycled by melting it down and using the steel to produce a later generation unit. On the other hand, energy resources, once used by conversion to thermal energy, not only cannot be recycled, but contribute to the imbalance of the system that is the beneficiary of their conversion.

Possibly the most valid comparison of energy resources to other natural resources is the comparison to air and water resources, insofar as the benefit to mankind is concerned. Few, if any, would disagree with the observation that man (or animal life of any kind) could not survive without air and water. The availability of air on earth is universal, and, contrary to energy resources, it is not considered in any way an article of commerce or an ingredient in the economic formula. Yet air, left to the ecological cycles of nature, is replenishing. The fact that we use it to sustain life does not consume it or use it up. It is only the excessive, nonbiological energy-consuming machines and chemical processes (motivated by energy conversion systems) that tend to degrade the air into a nonusable state. Water, the other resource without which life could not be sustained, is similar in many respects to air. There is, like air and fossil energy resources, only a fixed amount on the earth. However, like air, water is a recycling commodity in the natural order of things. If mankind utilizes water for biological purposes, it recycles and is available for use again at a later time, which time depends little on whether or not it was used. Like air, however, when one departs from the ecological cycle of use and injects into the water a chemical substance that nature has not provided for, the recycle rate is retarded, though the water is still not consumed.

Air and water are quite similar

From the standpoint of these self-recycling characteristics, air and water are quite similar. The fundamental difference is availability. Air is universally available. Water, however, is quite regional and local in its availability, which has had a striking influence on the pattern of development of civilizations. Early man made his home near freshwater streams,

lakes, and other natural resources. Today, virtually all large areas of population are located on major waterways, which provide for both transportation and water to sustain life. In areas in which water is available beneath the earth's surface, energy in some form is used to raise it to the surface. The local availability of water makes it a good deal different from air in the economic structure. Since all people do not have it available at their point of need, we accept the obligation to pay for it as a utility, recognizing that we are paying the cost of treatment and delivery.

Energy resources differ in two ways from air and water. First, without air and water, animal life could not exist under any circumstances, whereas man could survive in primitive fashion without consuming depletable energy resources. We have however, passed the point of no return in this regard. Mankind could survive in limited regions of the world on the daily energy from the sun, either as a direct form of heat or through the photosynthetic processes of plant life. But ever since the time that man learned to convert the stored solar energy in plant life to fire, the dependence on fire began, and now our entire modern civilization is based upon it.

The second fundamental difference is that, unlike air and water sources, fossil energy resources are not ecologically recyclable. Once the fossil energy has been extracted from the earth and converted to a useful form, starting with heat, nonusable chemicals are formed that will not return to a convertible energy resource.

It is the similarity of energy resources to air and water, coupled with the differences, that makes energy a truly unique commodity.

18
Energy transportation

A fundamental study of the phenomena of energy transportation is an essential ingredient to any comprehensive development of future energy policies, the optimization of energy-related parameters in end use systems, or sources to serve these systems. Energy transportation has been utilized for centuries, but previous developments have been the result of an immediately available supply to satisfy an immediately available demand. In many cases, the transport phenomena have been confused with the terminal use requirement or the source form. This chapter simply presents an overview of the concept of energy transportation as food for thought.

The first category to be considered is that of transporting the energy from its source through various stages of processing or conversion and then to the consumer or user. Only the current state of the art will be discussed initially. This category includes:

- Bulk or batch transportation of raw materials, such as coal ore and uranium ore; crude petroleum liquids or gases; processed solid fossil fuels; refined nuclear fuel; various grades of liquid petroleum products; gaseous fuels liquefied for transportation effectiveness or efficiency; gaseous fuels in gaseous phase; dry wastes for processing as fuel; and fuel processed from dry wastes.
- Electromagnetic wave transportation of solar energy to the earth's environment.
- Steady-state conduit flow (pipeline) of crude petroleum products to points of refinement or use.
- Pipeline flow of refined petroleum products to either distribution stations for transfer to other transport systems or into distribution piping systems and processed natural gas to distribution system storage and piping systems.

- Distribution piping system transport from storage or primary pipelines to user entities and distribution of various forms of manufactured gas to user entities.
- Transportation of electric energy from generating stations either directly to user entities or to use entities via network transport systems.
- Two-phase transport fluid systems, such as steam systems, that convey high-level thermal energy from a central conversion plant to user entities or refrigerant systems that convey low-level thermal energy from a user entity to a refrigeration plant.
- Single-phase transport fluid systems, such as high-temperature water or other thermal fluid, that convey high-level thermal energy to a user entity or low-temperature chilled water or other thermal fluid that conveys low-level energy from the user entity to a refrigeration plant.

User entity

The process needs or thermal environmental devices within the user entity (building or plant) can use a variety of energy forms. Also, the transport systems available to the site generally provide a choice of the forms available to satisfy the user needs. Usually, all the various process needs within an entity are not best satisfied by the same form of energy, and the same available form is not necessarily best suited to all the various process needs. Thus, based on the relevant parameters, a system designer or analyst must determine the form best suited to the needs of each individual process, the form(s) of energy available for purchase that could best serve the needs, the proper conversion system(s) to employ, and the method(s) of transport or distribution that is most effective between the point of entrance to conversion and, subsequently, from the point of conversion to the product. These sub-system parameters, although intimately inter-related, should be considered on the basis of individual merit.

The transport systems available within a building include:

- Single-phase high-level distribution, such as hot water or thermal fluid.
- Two-phase high-level distribution, such as steam.
- Single-phase low-grade high-level transport systems, such as condenser water systems or energy transfer systems for heat pumps.
- Two-phase low-grade high-level transport systems, such as refrigerant condenser circuits—hot gas and liquid.
- Single-phase low-level distribution systems, such as chilled water or thermal fluid.
- Two-phase low-level distribution, such as refrigerant suction vapor.
- Electric energy distribution via electrical conductors.

This chapter does not include a complete discussion on the relative features of each of these alternatives. However, the systems designer should thoroughly investigate the advantages and disadvantages of each of these distribution forms. Basic considerations should include: parameters of investment cost of both the distribution system and the interface at the process or terminal and at the source, flexibility for changes in source form, distribution system efficiency or energy effectiveness, maintenance requirements (and costs), and adaptability to future system growth or modification.

Although the energy transportation systems *to* the user entity are not of immediate concern to the systems designer, they are of paramount importance to society and its business institutions in the future. The ramifications of energy transportation must be considered in depth by corporate decision makers in both energy industries and the user facility industries and by policymakers at all levels. Energy is a resource; most forms are both depleting and highly localized in nature. Thus, the transport concept is the key to future viability of both local and national economies.

Perhaps it is the recognition by many that the universal distribution of solar energy directly to the user entity negates the need to address this complex problem and is thus responsible for the appeal of solar energy. Others have recognized the limitations of solar availability due to limited flux density and, consequently, have turned to the conceptual development of such transport systems as laser beams and extremely high-density electromagnetic radiation to address our transportation needs of the future. Until such times as these are developed to the demonstration stage, however, our needs must be satisfied by optimum utilization of current technology in conversion and transport.

19
Infinite source

In the engineering science, the fundamental textbooks have traditionally used the technique of reducing seemingly complex problems to a manageable level by removing all contributing parameters possible. Then, within the simplified scope, the solutions to the initial complex problem could commence one elementary step at a time. Examples of some such simplifications are free body diagrams, the black box concept, definitions of boundaries, and the concept of *infinite source-infinite sink*.

Unfortunately, in some disciplines of engineering, the purpose of the simplification has been lost, and the problem never returned to the initial integrated concept. (It might be mentioned here that the same criticism applies to many disciplines of the social sciences.) In the field of elementary thermodynamics, the concept of infinite source-infinite sink has been employed for decades as a very useful tool. However, as the engineering students entered the community of designers of energy conversion systems, the only areas in which the finite nature of any problem were reintroduced was that of economics or immediate availability.

The proposition is then presented that the lack of recognition of these sources and sinks as being finite by natural laws has been a significant contribution to two major sociotechnological problems of recent years: environmental degradation and energy excesses.

Sun is only infinite source

It is interesting to note the overwhelming attention given by the scientific community and the public sector to the perpetuation of the infinite source concept in looking to the sun for the long-range solutions to energy

problems. For, in reality, the sun is the only infinite source of energy available to mankind. The reasoning is, then, that if we simply develop the technology of converting the thermal solar energy into whatever form we desire (shaft, electrical, heat, or heat removal), we can proceed with business as usual. The fact is that the flux density of the solar energy is so low that the capital burden of collecting, storing, converting the form, and transmitting the energy obtained from our infinite source is so great that business as usual is not possible if this is the sole source available.

Much attention, primarily in government-supported research and development, has been directed to such schemes as the solar furnace (capable of generating fluids at a temperature level compatible with the state of the art methods for producing shaft power), oceanographic thermal level power plants, satellite converters, and collection by photosynthesis to grow gigantic forests for ultimate conversion to methanol. If one carefully scrutinizes the capital and labor, for both investment and maintenance, required to provide an energy unit equivalent to a single gallon of gasoline, or energy to "power" a home with the amount of electricity American homes consume today, it is seen immediately that, when this becomes our only refuge, it will not be "business as usual."

Finite source must be found

The purpose of addressing this issue is not to discourage the ever-important consideration of looking to the infinite source as an eventual means of providing some of our energy requirements, if not all; but rather to emphasize that the more immediate task to be addressed is the "finite source."

As the engineering, architectural, business,

and academic communities move aggressively in the direction of the finite source, our society will revel in change. The challenge to the architect to produce an esthetically attractive building that is functionally successful and utilizes the sun and environment in such a way as to minimize energy required for indoor comfort; the challenge to the engineer to provide the product requirements for such a building or for a vehicle with a system that is both capital effective and has minimal process losses; the challenge to the entrepreneur to develop and market products for a growing energy-conscious public; the challenge to educators to lead the way in this new sociotechnological era, probably the greatest challenge in the field of education since higher level education was made available to all who desired it; and the reward to homeowners who can maintain a standard of living not hooked on consumption of increasing amounts of energy; all will result from the hopeful recognition of the "finite source."

The point is not too subtle: this "finite source" concept is simply another approach to the almost worn-out topic of energy conservation. The need for the alternate approach is mandated by the lack of success of less fundamental avenues taken heretofore.

In the ensuing years, the accomplishment of the goals of the finite source concept will necessarily cause meaningful shifts in current economic and business institutions. There is little doubt that the revenue growth rate of energy suppliers will suffer, perhaps even assume a downward trend. This will require skilled management on the part of these institutions, management which must turn to development of infinite source energy *where it is most economically adaptable*, effective utilization of the finite source, and other such diversifications. Logically, an infinite reliance on a finite source is inevitably short-lived.

In summary, if the finite source thrust is successful, an increasing dependence upon the infinite source will become ever more practical. If only there were such a patent solution to the infinite sink problem!

20

An energy resource standard

Section 12 of ASHRAE Standard 90-75 provides the analyst or building owner with a methodology for estimating the quantity and forms of energy resources required to support new building operation. This "crystal ball" also provides an awareness of the energy dependence and costs of operating a facility.

Proposed Section 12 of Standard 90-75, "Annual Fuel and Energy Resource Determination," has been subjected to its second open review in accordance with consensus standards procedures. Panel 12 of the Standard 90 Project Committee—responsible for drafting, managing reviews, and achieving consensus—has completed a revised draft following resolution of differences and completion of revisions mandated by the comments on the second open review. The Panel submitted a recommendation to the Coordinating Committee of the 90-75 Project Committee indicating that the consensus had been achieved, and that Section 12 should be added to the Standard.

Section 12 is in a rather unique situation

This chapter was reprinted from *ASHRAE Journal*, May 1977 and was originally entitled, "Section 12: Toward a More Effective Use of Energy Resources." Appearance of this material in *Energy Engineering and Management for Building Systems* does not necessarily suggest or signify endorsement by the American Society of Heating, Refrigerating or Air-Conditioning Engineers, Inc. A copy of ASHRAE Standard 90, "Energy Conservation in New Building Design," can be obtained by writing to The American Society of Heating, Refrigerating and Air-Conditioning Engineers.

insofar as Standard 90 or any other consensus standard adaptation procedure is concerned: It is one section that is being reviewed separately from the complementary text of the Standard as a whole.

To inform readers of its status, a brief overview of the genesis and purpose of Section 12 appears useful.

During the original open reviews of ASHRAE Standard 90-P (preceding adoption of 90-75), numerous comments regarding the energy resources issue were submitted. The original eleven sections of the standard did not address this issue, and a summary of those comments was that if the Standard did not recognize energy resources, it was not addressing the fundamental issue of energy conservation; i.e., conservation of depletable or nonrenewable energy resources.

The Society recognized the legitimacy of these comments, and further recognized that inclusion of resource consideration would delay adoption of a much needed energy conservation standard for new buildings while the issue was being subjected to the needed consensus procedures. It was decided that

ASHRAE would adopt the Standard as originally conceived, omitting the energy resource, and issue a "Foreword" statement recognizing the need for considering energy resource and committing the Society to including this consideration as quickly as a text relating to this issue could be developed within the procedures established by ASHRAE and the American National Standards Institute (ANSI).

The original efforts in addressing the issue were made by an ASHRAE Presidential Committee which recommended that the original eleven sections of Standard 90 not be changed, except as required by other than resource issue comments; and that the entire resource issue be added as a separate Section "12." The Committee felt that this would enable the Standard to be adopted at the earliest possible date and would not create the need for significant changes to those sections.

As an outgrowth of this recommendation, Panel 12 was established for the purpose of drafting the resource section, conducting the open reviews, and obtaining the necessary consensus to add this Section to the Standard.

Shortly after adoption of Standard 90, the Panel drafted a text for Section 12 which was published in the July 1975 issue of *ASHRAE Journal*. The Project Committee felt that revisions mandated by the reviewers were significant enough to require a second review. The revisions required during the second review have been made and the Panel has recommended to the Standard 90-75 Project Committee that consensus has been achieved and that Section 12 be added to the Standard.

In the process of communicating with reviewers of Section 12, there appears to be one salient point of misunderstanding which might be addressed in some detail. If it can be said that there was any disagreement between the commentators and the Panel who drafted the text, it would be that the reviewers who perhaps remained "unresolved" in the consensus procedure did not recognize the value of resource consumption information to the systems analyst or building owner.

Section 12 provides the only available standard methodology for estimating the quantity and forms of energy resource(s) that will be required to support the building operation. Without the information developed in Section 12, the energy source and consumption information available to the analyst is left at monetary considerations and an arbitrary energy evaluation unit of British thermal units or kilowatt-hours. The latter has significance only if an arbitrary potential legislative limit is placed upon the building end use energy consumption, and the former has no relation to energy economics except for the obvious cost of the commodity.

Energy use can be calculated

Fundamentally, Section 12 provides a methodology whereby the systems designer can calculate the annual building energy consumption in the form consumed (MCF of natural gas, kW-hr of electricity, gallons of oil, etc.); then, by use of resource utilization factors (RUF), extend this information to the quantity and form of resource required to provide for the end use forms. The end result of the use of Section 12 is that the analyst or the building owner has complete knowledge of how much resource the proposed building (with its energy systems) will consume, and what form of resource will be required to support the building's energy system.

Section 12 does not mandate an optimization analysis to reduce the energy resource consumption to its minimum value. The logic behind this is that several resource options are available at most building sites, and the "reduction to minimum" approach may not necessarily be in the best financial interest of the building owner or in the best long-range interest of resource energy conservation. Buildings are not like vehicles, in that they have available to them alternative resource energy selections and are relative long-range investments. Thus, with lack of resource information, the building owner may elect to minimize the end use form and/or cost and later find his long-range dependence upon a relatively scarce source. In this regard, the use of Section 12 provides a "crystal ball" for the building owner and at the same time provides him with an awareness of the energy dependence and energy costs of operating the facility. Its existence serves as the only available nontrade source of knowledge for the building design profession regarding resource dependence.

Value of energy answered

Section 12 has achieved another significant milestone in the overall problem of energy economics: It recognizes the need for establishing a *value* of energy resources. This is addressed in Section 12.2, where the need for a resource impact factor (RIF) is implied. The output information developed in Section 12 is in the format that could be utilized in conjunction with the proposed RIF to optimize the energy source selection for a building project. The RIF concept encompasses such considerations as "availability, social, economic, environmental, and national interest issues." Until these are available, however, they remain the only judgment factor if the utilization factors of Section 12 are applied!

In conclusion, the singular question which welded the majority of the critical comments received in the Section 12 review was: "What will be the benefit of the use of Section 12?" The answer is that it will provide both analysts and owners with information currently unavailable to them. With this knowledge, decisions regarding energy resource and consumption will be founded on fact and will result in more effective use of energy resources. Additionally, Section 12 provides the basis for and recognition of RIF use and development—a concept that will ultimately facilitate intelligent national policy in building systems energy use.

ASHRAE is in a unique position in addressing these complex energy phenomena, in that it is neither a political body, nor a trade organization. To maintain this posture as a technical engineering society with the sole purpose of assimilation of technical information, it is imperative that they not yield to political or institutional pressures or submit to compromises to achieve their ends. They must, as a society maintain technical integrity in every undertaking including standards.

Standard 90 is perhaps the most ambitious standard that any technical society has undertaken, in that it is of significant interest as well to the public, institutional, and political sectors. As such, there has been significant pressure from these sectors to influence the technical content of the document. It is the Society's obligation to recognize and consider these inputs and to maintain the technical integrity of their efforts. In the adoption of Section 12, the Panel kept this goal in sight at all times and hopefully has preserved, on behalf of ASHRAE, this technical integrity.

SECTION III

Energy economics

The topical discussion for the preceding section introduced the concept of an energy revolution. That concept is addressed in a constructive manner in this section wherein energy economics is suggested as a tool for molding the direction in which the energy revolution takes us. The first two chapters in this section present the proposal that a science of energy economics, apart from but intimately related to monetary economics, be developed. This sudden need for a "new science" should be considered in perspective.

In the early days of the industrial revolution, the physical laws relating to the engineering sciences were not nearly as well understood as they are today. Furthermore, the physical properties of materials were little understood and good quality control for processing these materials did not exist. This lack of knowledge did not prevent the development of the early mechanisms and machines. The designers and inventors developed the designs much by trial and error. Starting with an idea, a device would be constructed. If it worked, that was fine; if it failed, the cause of failure would be sought and another attempt made. It was through this series of experiments that the physical sciences and knowledge of materials that exist today were developed. The important point is that the early need for knowledge was recognized by the scientific leaders of the time, and they set about a course of developing the science.

An example might be the fascinating development of thermodynamics in the nineteenth century, i.e., a system of laws and principles that were developed and pieced together in such a way that we now have a thoroughly organized science that is universally taught to engineering students, who learn and accept the complex laws almost as a second nature.

Energy, although an essential ingredient in the development of machines and societies, was never considered to relate to the overall problem in as complex a form as the laws of thermodynamics. It was an ingredient that was more or less taken for granted—either it was available or it was not. In the Alpine areas, electricity made from hydroelectric plants is the primary energy form. In central Europe and northern England the primary energy form is coal. As the use-rate curve becomes steeper, and mankind starts to realize the finite nature of the current source, it becomes evident that the only way the future can be secured is through some efforts at planning. In a radio address in 1932, Franklin D. Roosevelt said, "These . . . times call for the building of plans. . . ." The time for disorganized trial and error in the depletion of energy resources must come to an end or our present social systems will certainly collapse. Although we may lack many of the answers, as the engineers of the latter eighteenth century lacked knowledge of thermodynamic laws and materials, we can at least initiate the organization of the science. If such a proposed science can germinate, there is little doubt that the knowledge will follow.

The alternative is that the present situation will continue. This situation is one of total polarization between the disciplines knowledgeable in the energy situation. As examples, physicists are in diametric disagreement on the issue of the present wisdom of and future place of nuclear energy; the solar utilization concept has developed into camps of proponents and opponents in which the scientific facts are little discussed; the United States federal government attempts to get the public to reduce consumption while the industries that produce and sell energy continually attempt to encourage more use. The economists insist that energy is merely another commodity to be thrown into the hopper with whatever else happens to appear on the commodity market while the engineers are torn between the logic of designing machines for most effective utilization of energy and the pressures to give paramount consideration to manipulated monetary structures.

The chapters in this section, then, are intended to explore the need for such a proposed science of energy economics, recognize some of the essential elements in such a science, and reveal some of the challenges to be addressed by the science. There is little doubt that the engineering community properly motivated can contribute significantly to the overall solution of the energy problem. An example of energy waste resulting from the failure to consider energy apart from its monetary value is developed in some detail in the chapter on the "Case History Study" and the affect of a misdirection in engineering philosophy is revealed in the chapter "Infinite Sink!" The discussion of second law concepts reveals some of the challenges which the engineering community may address. Indeed, "these times call for the building of plans!"

21
Energy economics is a needed science

Mechanical engineering has classically been defined as "the applied science of energy conversion." Thus, it is somewhat of a contradiction for practitioners in the HVAC field (a subscience of mechanical engineering) to undertake the goal of "energy conservation." The practice of minimizing energy consumption in building systems, whether in design or operation, might therefore be more accurately described in other terms.

After some reflection and research, we suggest that *energy economics* is an appropriate label to attach to such efforts.

Webster defines *economics* as "a social science concerned chiefly with description and analysis of the production, distribution, and consumption of goods and services." In the past, most efforts involving energy consumption by the building systems design profession were, understandably, motivated by monetary economics. And though it is certain that monetary economics will continue to be a primary design parameter, it is becoming increasingly evident that another parameter will have an even more meaningful effect on both system design and energy source selection—the parameter of pure energy economics. Examples of this include ASHRAE Standard 90, the Federal Emergency Building Temperature Restrictions of 1979, the Federal Building Energy Performance Standards, and many other proposed and adopted national, state, and local statutes.

Thus, while we cannot omit monetary economics from our considerations, we must supplement it with a new science of energy economics. At the outset, let us make two basic observations:

• In the definition of *economics*, no limitations are placed on "description and analysis" in monetary terms.

• Energy, as obtained in the world today in its initial or raw material forms, can be considered "goods" in the light of Webster's definition.

After accepting this, we see that the basic hypothesis underlying further development of the independent science of energy economics is:

The forces that motivate and strengthen the monetary economics of a society are diametrically opposed to the conservation of energy resources.

In any socioeconomic system the increased use of energy serves to strengthen the so-called economy of the society. Inversely, conserving the energy resources has the effect of retarding economic growth and weakening the economy.

Since none of the commercially available energy sources today are replenishable, however, continued accelerated use of these commodities is tantamount to living on credit; i.e., today's use, and the advantages thereof, will not affect society until tomorrow. This observation leads to the thought that enters the mind of every intelligent and decision-making businessman and homemaker—defining the risk. The risk of using tomorrow's resources today must be taken in a calculated manner; or as the conservative gambler might say, we have to hedge our bet. We must always be able to retain the flexibility to regroup and change the game plan when the need arises.

It is with this concept that the economics of

energy must be disassociated from classical monetary economics. In the development of the proposed science, the basic laws of economics can be utilized, keeping in mind the fundamental difference that money is generated by consumption of resources whereas energy use depletes our finite resources. Thus, the basic economic law of supply and demand is the irrefutable footing of the science of energy economics.

Upon this footing can be constructed the science: The supply is our limited resources, and the demand is what a sound monetary economic system must have to survive and grow. This balance between monetary economics and energy economics must be defined and then achieved on national levels (if not worldwide) if mankind and nations are to survive.

The science of energy economics must recognize that technical breakthroughs occur at (hopefully) opportune times in the history of civilization and provide for these eventualities. We should never be in the position of betting all we have on such an eventuality, however. For example, in the 1940s and early 1950s, we rested comfortably in the thought that nuclear energy represented a new found infinite source. Now we are beginning to realize that the loop is not closed; we still lack a method of disposing of the wastes.

In summary, a new subscience of energy economics is needed as an evaluation tool for professionals in energy conversion as well as for society in general. This science can use as its footing the basic law of supply and demand and build upon it in much the same manner that monetary economics has developed. In its application, energy economics should interface with monetary economics as well as affording an independent input parameter to the design of all energy conversion cycles, systems, and machinery. The science should be established on a worldwide basis, but until current national boundaries are transcended by interaction among governments and peoples, it must be applied by each nation individually for its survival and defense.

22
The energy hypothesis

In Chapter 21 the following hypothesis was introduced:

"The forces that motivate and strengthen the monetary economics of a society are diametrically opposed to the conservation of energy resources."

This chapter discusses the significance of this hypothesis as it relates to the energy economics efforts of the engineering community and governmental policies.

The statement was called a hypothesis, not a law, on the basis of the definition of a hypothesis: an interpretation of a practical situation or condition taken as the ground for action.

Examples that justify the hypothesis are endless. To state a few:

• The monetary economics of the energy industries are such that retardation of ever-increasing energy use could financially destroy them. Consider the commonplace utility rate structures that penalize a consumer for not using more hours (energy) of demand (power); or the rates for either fossil fuels or electric energy, which reduce the unit cost and consequently the *average cost per energy unit as consumption increases*.

• Examples in transportation are numerous. Consider the complex interrelations of the automotive industry, the highway systems, and the building industry. Development of interstate highways into and around major metropolitan areas spurred development of suburban or satellite business communities such as office and industrial parks as well as housing communities and associated commercial developments. The only links between the housing communities, suburban business communities, and "downtown" areas are the super-

highways. The required use of the family car for transportation to a daily place of work not only increased the per capita gasoline consumption significantly, but also increased demand for the second car, which was now sorely needed for visits to the doctor, trips to the store, and shuttling the kids to their extracurricular activities. We became a nation on wheels with no alternative. Yet, no one lost in this conversion: the building industry boomed; the automobile business boomed; and the energy business boomed.

• An example that presents an overview of the hypothesis is shown in Fig. 22-1. Against the time span of 1909 to 1973 is plotted gross national product per capita in units of thousands of dollars; superimposed is the annual energy consumption per capita in units of billions of Btus. It is immediately evident that the two curves are very nearly congruent. This phenomenon tends to prove the legitimacy of the hypothesis.

Several conclusions could conceivably be drawn from the above examples, particularly the one illustrated by the graph, and some of these conclusions are most distressing. For example, one might conclude that the two phenomena are so closely interrelated that by some law of economics a downward slope in one would dictate a downward slope in the other, and vice versa. This conclusion could likely be justified by the current situation in the United States. We have been effectively reducing our energy consumption by various external and internal pressures; and concurrently, we are in the midst of one of the most complex monetary crises of modern times—

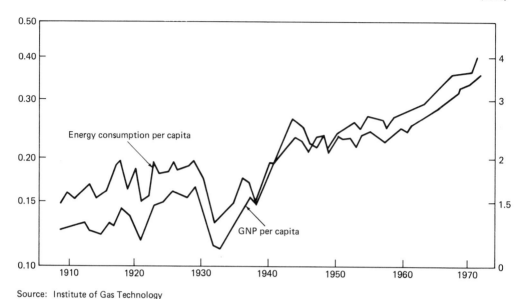

Source: Institute of Gas Technology

Fig. 22-1. Trends of per capita GNP and annual per capital energy consumption in the U.S. 1909–1973.

unemployment is high, inflation is rampant, shortages prevail, and many of our largest business institutions are in serious financial difficulty.

Another interesting observation is that previously all discontinuities or changes in direction of the curves were caused by other forces affecting GNP, with the energy curve following. But this time, the energy curve is the controlling parameter, and the GNP curve is doing the following. This simple observation, if studied in more depth, could go a long way toward explaining the current economic situation.

But in the study of economics, the laws are not as clearly defined or as hard and fast as the laws of the natural sciences; and in most cases, the recognition of a phenomenon as stated in the hypothesis can provide guidance for intelligent solutions to seemingly overwhelming problems. If, for example, in the development of national energy policies, every effort is made to defy the hypothesis, to separate the two curves—bending the energy curve horizontally if not downward while forcing the GNP curve upward—further crises can be avoided.

There is no question but that this can be

achieved, and this is the goal of energy economics. In the design of any energy conversion system or the conception of any alternatives to existing methods, the use of energy economics should include the identification of the "energy product" requirement. This identification should start with the concept and be carried through the entire design development of each and every energy related component.

For example, in transportation the "energy product" is simply the theoretical units of energy required to move an individual or given cargo from Point A to Point B in a given time. From this point, the burdens of the practical systems such as vehicles are considered. The phase-by-phase degradation continues until we have achieved a method of providing for the "energy product" in the most effective manner. The final measure is to compare the resource utilized to achieve the product requirement. This evaluation parameter is commonly identified in the automotive area as miles per gallon or in public transportation as passenger miles per gallon.

The same approach can be validly applied to building systems, industrial processes, etc. In Section 12 of ASHRAE Standard 90, this concept is applied to a degree to building systems.

In the development of energy policies, whether they be national, international, or regional, consideration must be given to the energy hypothesis. The question must be answered: Will the policy reduce energy use while not adversely affecting the monetary economic situation? There are numerous ways of achieving this goal either totally or in part. Lack of recognition of this interplay has plagued us so far.

23

A case history study illustrating the need for energy economics in design

A comprehensive study of campus cooling points up the necessity of incorporating the energy economics parameter into the design of building systems, and specific recommendations tell how to go about it.

Chapter 48 entitled "Integrated Loop System For Campus Cooling" presents the background of a study that resulted in the recommendation to install an integrated loop chilled water system on the campus of the University of Missouri–Rolla.

Briefly, the recommendation called for existing chilled water plants serving individual buildings to be connected in a loop system. Because of system load diversity, they would thus be able to provide chilled water for additional buildings currently served by a variety of packaged equipment and window air conditioning units (see Fig. 23-1). This would be accomplished with significant savings in both energy consumption and machine operating hours as compared with the installation of separate central systems in the buildings now served by unitary equipment.

In the process of gathering the information needed to make this study (heating and cooling loads, types and sizes of refrigeration machinery and auxiliaries, types of building environmental systems, etc.), significant data were developed. These data serve as the basis of this chapter—a case history report on building energy economics comparison. The information is presented to accent the differences in energy demands and consumptions among various buildings with similar uses and occupancy schedules. The excesses of energy required by some indicate that although the buildings were all relatively successful designs architecturally and environmentally, energy economics was not considered as a design parameter.

This is not a unique situation. For the past three and one-half decades in the United States, the building environmental sciences have been undergoing probably a more dynamic progression than any other field of engineering in history, save space technology. As a result, design practitioners have had all they could cope with in keeping pace with progress in building materials, building systems, fenestrations, mechanical system concepts, and new machinery. In addition, they have had to meet the demands of the marketplace from the standpoint of ever-increasing requirements for multiple control zones. The additional pressure of satisfying these demands within the budget restrictions established for earlier, less sophisticated systems posed a challenge of almost astronomic proportions to the systems design profession. With these burdens, and with readily available

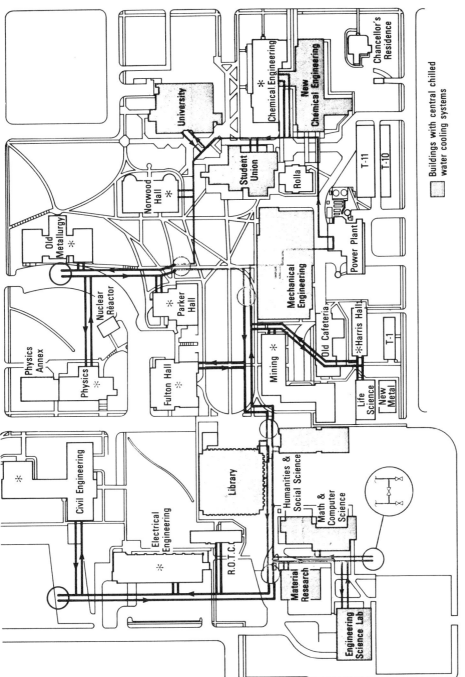

Fig. 23-1. Campus of the University of Missouri-Rolla. Buildings shown in color have central chilled water refrigeration plants. Colored and black piping together represent the proposed integrated loop system described last month. Phased construction was recommended, the piping shown in black representing the last stage of loop development pending further campus expansion and the circles on the piping denoting connections to be made as the loop is developed and expanded. Buildings marked with asterisks are those without central cooling, considered for inclusion in the integrated loop system.

Legend: ▨ Buildings with central chilled water cooling systems and competitively priced energy sources, design practitioners understandably did not consider energy economics as a design parameter.

It might be emphasized at this point that building environmental systems include lighting as well as HVAC systems. Thus, a major portion of the total output of electric power generating plants serves building environmental needs. Other major areas of energy consumption are the categories of transportation and industrial processes, and neglect of energy economics is as manifest in these areas as it has been in the building industry.

Building loads compared

The Rolla campus includes ten buildings with central chilled water systems. Table 23-1 shows the comparative loadings of the ten buildings. They are seen to range from 147 sq ft per ton (82 Btuh per sq ft) for the Student Union to 594 sq ft per ton (20.2 Btuh per sq ft) for the Library. This comparison, however, is not especially revealing since occupant loading densities and functions are considerably different for these two buildings. If these two buildings and the University Center are discarded, the remaining buildings are similar in function, occupancy densities, and occupancy schedules. These range from 194 sq ft per ton (62 Btuh per sq ft) for the Mechanical Engineering Building to 413 sq ft per ton (29 Btuh per sq ft) for the Math & Computer Science Building.

An understanding of the differences in specific capacity requirements is essential if one is to make use of this information in the development of methods for minimizing energy usage. Although specific capacity is only one area of concern, it is the area that often is completely out of the control of the mechanical systems designer (except for the ventilation rate contribution). This specific loading is the calculated coincident full load that the building (plus ventilation) imposes on the system. Not included in this form are any inherent system inefficiencies, such as reheat, refrigeration and heat energy in reduced load zones at the time of maximum building coincident load, distribution system losses, etc.

Lighting levels range from 1.63 W per sq ft for the Library to 3.64 W per sq ft for the Materials Research Building. (The values are based on total installed lighting per gross area.) Again, if the nonsimilar use buildings are discarded, the range is 2.44 W per sq ft for the New Chemical Building to 3.64 W per sq ft for the Materials Research Building. The correlation between lighting power density and specific cooling load, however, while not overly impressive, is significant; a calculation for the seven similar use buildings reveals a positive correlation of 0.4. This is not surpris-

Table 23-1. Calculated cooling loads versus installed capacities for buildings with central chilled water plants.

Building	Area, sq ft	Cooling load, tons	Installed capacity, tons	Load, sq ft per ton
Physics Annex	14,800	37	71	400
New Chemical Engineering	78,600	268	360	293
Library	85,600	144	274	594
Mechanical Engineering	38,922	200	200	194
Student Union	17,900	122	115	147
University Center	38,400	166	204	231
Materials Research	28,600	101	120	283
Math & Computer Science	35,900	87	123	413
Humanities & Social Science	30,600	123	155	249
Engineering Science Lab	42,400	155	195	274
Totals/average	411,722	1403	1817	293

ing since any building design feature that adds to the cooling load should obviously be scrutinized if energy economics are being considered.

Another feature that has an appreciable effect on cooling load but was rejected from a quantitative analysis is the fenestration of the different buildings. A correlation between fenestration and specific cooling load for these buildings was similar to the lighting level correlation ($r = 0.36$). But since all relevant variables were not considered (such as reflectivity of glass, interior shading methods, and exterior shading versus orientation), this feature was not included.

Effect of auxiliaries

Thorough understanding of the building features that contribute to specific cooling load is the very starting point in designing with energy economics as a parameter. If the total cooling load increases, the base power requirement to drive compressors (W) or absorption machines (Btuh) increases, the power requirement to drive all necessary auxiliaries increases, and the resulting energy consumption increases. As stated previously, the Mechanical Engineering Building, at 62 Btuh per sq ft,

has more than twice the specific cooling load of the Math & Computer Science Building, at 29 Btuh per sq ft.

The mechanical systems designer is seldom in a position to control the specific cooling load, a topic that will be discussed again in a subsequent section. He is in sole control of the energy usage per unit of building requirements, however. Specific system power requirements, SSP, is defined as the power per unit of cooling capacity as required by system design, expressed in such units as Btuh per ton or kW per ton. Specific system energy, SSE, is the seasonal or annual energy consumption per unit of system capacity requirement, expressed in such units as Btu per ton or kW-hr per ton.

All of the buildings studied have absorption chillers, except for the Student Union with an electric centrifugal compressor. Specific system power requirements and specific energy requirements were analyzed for the ten building systems, and Table 23-2 illustrates the comparative mechanical system power requirements. The refrigeration auxiliaries include chilled water pumps, condenser water pumps, cooling towers, absorption unit pumps, and control power. The fan auxiliaries

Table 23-2.   Comparative mechanical system power requirements.

Building	Area, sq ft	Installed capacity, tons	Fan auxiliary power kW	Refrigeration auxiliary power		Total auxiliary power		
				kW	kW per ton	kW	kW per ton	W per sq ft
Physics Annex	14,800	71	8.8	22.1	0.3112	30.9	0.4352	2.0878
New Chemical Engineering	78,600	360	67.5	80.0	0.2222	147.5	0.4097	1.8765
Library	85,600	274	84.0	115.5	0.4215	199.5	0.7281	2.3306
Mechanical Engineering	38,922	200	42.1	66.5	0.3325	108.6	0.5430	2.7901
Student Union	17,900	115	12.0	27.7	0.2408	39.7	0.3452	2.2178
University Center	38,400	204	27.0	52.4	0.2568	79.4	0.3892	2.0677
Materials Research	28,600	120	7.5	24.0	0.2000	31.5	0.2625	1.1013
Math & Computer Science	35,900	123	54.3	44.4	0.3609	98.7	0.8024	2.7493
Humanities & Social Science	30,600	155	23.1	28.1	0.1821	51.2	0.3303	1.6732
Engineering Science Lab	42,400	195	76.0	45.6	0.2338	121.6	0.6235	2.8679
Totals/averages	411,722	1817	402.3	506.3	0.2786	908.6	0.7787	2.2141

include only the supply and return air fans (fans necessary to effect space conditioning). These power loads are then brought to a specific requirement by referring to the common denominator of installed tons.

Figure 23-2 illustrates these data in bar graph form.

It is important to note that if these power units were brought to the base of kW per building ton, they would be higher in many cases since installed capacities exceed building loads, partly because of system parasitic loads (required for control purposes) and partly

because of available machine sizes. This is illustrated in the last column of Table 23-2, which expresses total system auxiliary power requirements per square foot of building area. (Note that this combines the specific load with specific power.) In this evaluation, building function does not have the relevance it did in comparing specific cooling loads, as long as the comparison is made in kW per system or installed ton.

Brief descriptions of building systems are given in Chapter 48 and will not be repeated here (but rudimentary information is given in

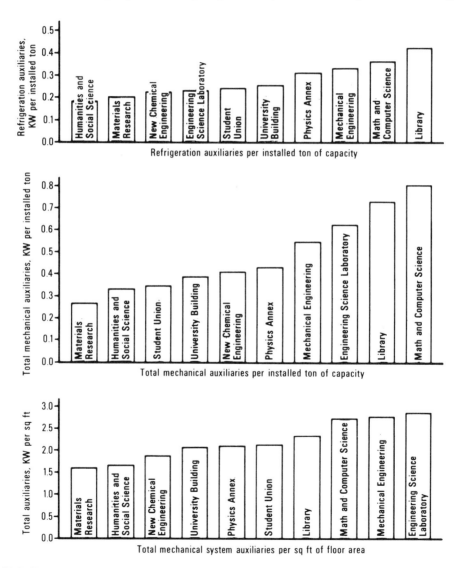

Fig. 23-2. Power inputs for mechanical system auxiliaries in the different buildings compared. Total mechanical auxiliaries include refrigeration auxiliaries and supply and return air fans.

Table 23-3). Before any value can be assigned to the data, however, one must understand the systems so as to determine the sources of deviations. Contributing factors included:

• Of the four buildings with the highest specific power requirements, three employed high-pressure air distribution systems.
• The cooling tower fan motors ranged from 0.065 to 0.275 hp per ton. The primary difference appeared to be higher horsepower requirements for less real estate. That is, when space limitations prevailed, power requirements increased.
• Chilled water pumps ranged from 0.02 to 0.21 kW per ton.
• Condenser water pumps ranged from 0.032 to 0.15 kW per ton.

Although the absolute values for the above specific requirements are small, they are significant in their deviations: 420 percent for cooling towers, 1050 percent for chilled water pumps, and 470 percent for condenser water pumps. When extended to system capacity, these deviations are *most* significant.

Annual energy consumption

Power, the time rate of energy production (or consumption), is a significant consideration in the cost of energy since power dictates the investment in generating plant and distribution system. When considering basic resources, however, the energy unit becomes the primary target.

Annual energy consumptions for the buildings were determined using a computerized calculation technique that has been verified in many buildings. The results are shown in Table 23-3, together with specific building energy requirements, SBE, for both electrical and thermal forms, and an identification of the type of terminal system control. (A combined result of specific energy and specific loading, each for the time period of one year, in units of KWH per sq ft and Btu per sq ft, is defined as the specific building energy consumption.)

Inspection of the results leads to the following observations:

• There is a positive correlation between

Table 23-3. Total annual electrical and thermal energy requirements of buildings studied. Aborption refrigeration is used in all buildings except the Student Union, which has an electric motor driven centrifugal.

Building	Area sq ft	System[1]	Annual energy consumptions				ΣSBE,[2] Btu per sq ft
			KWH × 10^{-3}	KWH per sq ft	Btu × 10^{-6}	Btu per sq ft	
Physics Annex	14,800	MZ	220.3	14.9	535	36,200	200,200
New Chemical Engineering	78,600	DD/VV	1201.1	15.3	3876	49,200	217,200
Library	85,600	DD	1590.2	18.6	3525	41,200	246,200
Mechanical Engineering	38,922	RH	622.8	16.0	3427	88,000	264,000
Student Union	17,900	HCO	348.5	19.5	106	59,200	273,400
University Center	38,400	MZ	681.1	17.7	4769	124,000	319,000
Materials Research	28,600	HCO	390.4	13.6	1250	43,700	193,000
Math & Computer Science	35,900	RH	749.7	20.9	5289	147,000	377,000
Humanities & Social Science	30,600	RH,HCO	439.3	14.3	4506	147,200	304,000
Engineering Science Lab	42,400	RH	971.9	23.0	7649	180,000	433,000

[1]MZ = multizone; DD = dual duct; DD/VV = dual duct with variable volume mixing boxes; RH = reheat; HCO = heat-cool-off.
[2]SBE = specific building energy consumption.

systems with higher specific electrical energy consumptions and those with higher specific power consumptions. (This correlation, though only on the order of 0.11, is positive and can be considered as having some significance.)

• Of the four buildings with the highest specific building electrical energy consumptions, three have high-pressure fan systems.

• *The building with the electric chiller does not have the highest specific building electrical energy consumption* (although it does rank third).

• The three similar use buildings with the highest specific building thermal energy consumptions all have reheat systems.

These results are not significantly different from what one would expect. The contribution of a high-pressure fan system to specific electrical energy consumption is obvious when one realizes that fan horsepower is directly proportional to pressure and that the fan operates for all the occupied hours of the building (i.e., it does not unload at reduced capacity, at least not in the systems installed in these buildings).

Analysis of a reheat system reveals that absolute energy consumption increases with decreasing building load. In dual stream and heatcool-off systems, on the other hand, energy consumption decreases with decreasing load. (With any energy conversion device, however, the consumption per unit of demand increases in an exponential manner with a decrease from full load.) Unfortunately, the humidity control capabilities of these three alternatives are inverse functions of their inherent system specific energy consumptions.

The last column in Table 23-3, titled summary specific energy (ΣSBE), is intended to illustrate specific energy consumption as a single quantity; it was developed using the arbitrary value of 11,000 Btu per delivered KWH for electrical energy. (The calculated thermal energy and electrical energy consumptions include all conversion losses and thus are the quantities delivered to the buildings.) From the standpoint of pure energy economics, this is the most relevant comparison. Since

the conversions are arbitrary and subject to challenge, however, they are presented simply for interest. Attempts are currently being made to develop acceptable data for extension of this unit in building energy economics studies.

Energy economics

The Rolla study, which shows a differential of 123 percent between the similar use building with the lowest summary specific building energy consumption and that with the highest, exemplifies the need to include energy economics as a parameter in building environmental system design. The following considerations are offered as a positive approach to incorporating energy economics into the design of all building systems.

• *Energy consciousness in building design* — This consideration is not normally under the direct control of the mechanical engineer on the building design team. Since it has such a direct effect on investment cost as well as on energy consumption of environmental systems, however, he has an obligation to make architects and electrical engineers aware of the implications of their designs with regard to the energy aspects of the building. As was emphasized earlier, it is at this stage of design development that the very significant specific building load, is established.

• *Intelligent definition of design parameters* — Available analytical data have been published for some time to illustrate the specific energy requirements of the three basic system control methods. At full cooling load conditions, all three might have the same specific energy requirements. But at reduced load conditions (which an analysis of any building system will show to exist during a majority of the operating hours), the requirements of these systems differ significantly. In order of increasing energy consumption at reduced loading, they are: heat-cool-off and variable volume, dual stream (double duct or multizone), and reheat.

Unfortunately, if one lists the same three basic systems in order of their ability to maintain space humidity control, the order is reversed. Thus, it is the responsibility of the systems designer considering energy economics to establish intelligently the *real* building requirements vis-a-vis humidity ranges or limits and then design a system to provide that level of control and no more. This approach is quite revealing, and it leads to the concept of incorporating energy units into the ASHRAE comfort chart.

Additionally, the single fan "economizer" type system must be scrutinized carefully in the process of systems selection. In many applications (those involving multizone or dual duct installations), specific energy consumption with such a system is higher than when some reduced capacity refrigeration machinery is operated in cold weather periods. The logic is quite obvious: with a single fan "economizer" system, the heating deck becomes a reheat device, and thermal energy consumption increases.

• *Detailed load calculation and part load analysis* — As stated previously, any energy conversion system is less efficient at reduced load operation than at design loading. Thus, if a system is sized to a "safe" load calculation, it will operate at a lower percentage of full load, and a lower efficiency, at all times. The latest computerized techniques of calculating heating and cooling loads should therefore be employed. As an example, if a system or component is oversized by 25 percent, at peak load it will never operate above 80 percent of design capacity, a point of higher specific energy consumption than at full capacity.

The part or reduced load analysis is a very tedious one to perform, but computer programs are available to assist designers in this effort. A part load analysis will reveal the hours per year or the percent of total operating hours that the system will "see" various increments of the full load. By using this analysis in selecting machine size increments, the designer can assure the greatest number of operating hours at the greatest machine load ratios, thereby greatly decreasing specific energy consumption.

• *Flow rate analysis* — Although two-phase steam systems are still widely used for primary heat transfer, and for good reason, the majority of systems employ the single-phase hydronic concept for terminal fluids. These systems use the concept of temperature differentials to achieve thermal capacities. In the design of any such system, the systems engineer should employ the maximum possible temperature differential. As an example, if a system temperature differential of 60 F can be achieved in a heating system, the water flow rates and consequently the pumping horsepower can be reduced to one-third of those for a system design based on a 20 F temperature differential. There are several techniques available for achieving this, among them coils designed for large drops and series connection of loads. Primary-secondary circuits are sometimes required, but the total pumping energy is still reduced.

It is more difficult to achieve larger temperature differentials with chilled water, but consider that a 15 F rise in lieu of 10 F will result in a 33 percent reduction in flows and, consequently, in energy expended for pumping. An additional benefit is usually lower investment cost in pumping and distribution systems.

• *Air system pressures and operating schedules* — Although some control schemes require high air pressures, the use of high-pressure high-velocity air distribution systems should be avoided unless techniques are employed to reduce fan energy consumption at reduced building loads. If a variable volume refinement is not incorporated, fan energy will often exceed the energy requirement of the basic refrigeration machinery because of the greater number of hours of fan operation.

An additional consideration in the selection of fan systems is the building occupancy schedule. The possibility of operating a central fan system a number of hours per year with greatly reduced occupancy rates should be prevented.

• *Primary conversion machinery selection* — Primary energy conversion apparatus such as boilers, chillers, and condensing units should be matched in size modules to the system's part load profile. Each module of machinery should be sized so that it will rarely operate at less than 50 percent of design or rated load. A study of most commercially available energy conversion systems will reveal that the energy input rate (per unit output) increases exponentially as the output decreases. The reason for this is the basic parasitic loading that contributes to mechanical inefficiencies at full load; most of these do not decrease as the output decreases, and with the loss remaining constant, apparatus efficiency decreases.

Capacity reduction modes constitute an additional consideration in the application of refrigeration machinery in smaller (reciprocating) sizes. Analogous to terminal system control, the capacity reduction mode that provides the best or most consistent result is the one that consumes the most energy. Compressor control modes, in order of increasing part load energy consumption, are: on-off, cylinder unloading, hot gas bypass.

• *Refrigeration and boiler system auxiliaries*—Energy consuming auxiliary devices such as condenser water pump motors, cooling tower fan drives, forced draft fan motors, feedwater pump drives, etc., should be scrutinized thoroughly to determine the combination of devices that imposes the lowest power burden on the system and thus results in minimum energy consumption. Cooling tower fan drives can be very low in specific power requirements, but selection without concern for this parameter commonly results in requirements as high as 0.25 kW per ton. In applying condenser water pumps, one must exercise care to arrange the tower, cold water sump, and pump to minimize static lift. In larger systems, multiple cells and variable pumping provide means of optimizing energy use effectiveness.

• *Energy conservation devices*—There are numerous devices and components available that have been specifically developed and marketed to conserve energy. Some examples are heat reclaim wheels, double bundle heat pumps, "heat of light" systems, so-called total energy systems, water source incremental heat pumps, etc. These devices cannot be overlooked in any system design; they must be considered. Unlike the measures recommended in the previous considerations, use of these devices may often penalize investment

cost, returning the increment out of operating energy cost savings.

• *Energy source selection*—The selection of an energy source or sources should be made independently of the above considerations. Once specific building energy consumption has been minimized, the source selection is relegated to owning and operating cost economics (present & anticipated) and availability trends.

Suggested checklist

Possibly, the major pitfall of designers has been to initiate an energy economics evaluation with a study for energy source selection, a technique that clearly is analogous to the tail wagging the dog. When the above considerations are applied to the energy economics parameter, power and energy consumption are minimized; and thus the cost will be minimal, regardless of the source. A suggested checklist for energy source selection is:

1) cost;
2) availability;
3) efficiency of conversion, at full and part load;
4) investment cost for storage, handling, and conversion apparatus;
5) environmental requirements of space;
6) environmental requirements of community;
7) demands and consumption of system;
8) availability of apparatus involved and its maintenance and service availability;
9) reliability of source;
10) reliability of conversion apparatus.

Owner's design guidelines

In applying the parameter of energy economics to building systems, the designer must take care not to introduce complexities that cannot be understood by the maintenance and operating staffs. Complete understanding is necessary if the design intent is to be carried out. Many efforts at achieving energy economics without consideration of this point have led to failure of a system to operate efficiently and, in many cases, failure even to satisfy performance requirements. Thus, the designer must always be guided by the rule that simplicity in design will result in successful performance. This consideration certainly affects the specific building energy requirement, and it offers more assurance that design performance will be achieved.

The University of Missouri-Rolla, as a result of the study, developed a set of system design guidelines to be applied to future buildings on campus. Those guidelines involved in energy considerations are:

• Whenever possible, cooling coil capacity control shall be by throttling valve, resulting in load response by variable flow.

• No loads that are humidity-critical, because of process requirements, shall be connected to the building chilled water system. All loads representing normal occupancy or human comfort shall be connected to the chilled water system.

• Special attention shall be given to minimizing all system auxiliary motor loads.

• If the following limitations are exceeded, special permission must be obtained from the University: refrigeration auxiliaries, 0.25 kW per installed ton; system auxiliaries, 0.25 kW per installed ton.

• Every effort shall be made in building and system design to minimize the energy requirements of the environmental systems. A thorough analysis of the energy requirements shall be performed and reviewed with University authorities prior to final design development. Areas of special attention shall include: lighting levels, light switching techniques, ventilation requirements, system selection and control logic, and fenestration.

These design guidelines could be considered a method of energy rationing, but it is hoped that they will simply serve to make design teams conscious of the energy economics parameter.

The author wishes to acknowledge Raymond A. Halbert of the University of Missouri and Victor E. Robeson of the University of Missouri-Rolla for their role in the initiation of this study.

24

Proposed format for organizing the study of building energy economics

. . . Because of the demands of an affluent society and readily available, competitively priced fuel sources, major efforts of the engineering community were concentrated on performance parameters rather than energy economics . . . A point has now been reached in our society where this concept of energy economics will soon be a lasting criteria in the design of any energy conversion system. Thus, a new subscience is imminently required— the applied science of energy economics . . .

Because of the revolutionary advances in building environmental technology of the last four decades, approximately one-third of all energy resources consumed today is converted directly or indirectly for the purpose of environmental control in building spaces. During this period of rapid technological growth, the demands of an affluent society and the ready availability of competitively priced fuel sources led the engineering community to focus its major efforts on performance parameters in preference to energy economics.

However, competitive market pressures

This chapter was reprinted from *ASHRAE Journal*, May 1974 and was originally entitled, "Applied Science of Energy Economics in Building Systems." Appearance of this material in *Energy Engineering and Management for Building Systems* does not necessarily suggest or signify endorsement by the American Society of Heating, Refrigerating or Air-Conditioning Engineers, Inc.

have recently led some practitioners to explore methods of evaluating the economics associated with a building's energy systems. One segment of the industry, system design, is concerned with the development of specific techniques for applying these evaluations. As it becomes increasingly evident that the world community is approaching the intersection on the curves between available energy resources and immediate demand, these techniques will provide the nucleus for a new parameter in building systems design: energy economics.

The concept of energy economics should be a primary criterion in the design of any energy conversion system. If it is to be useful to more than a limited number of informed specialists, its guidelines must be based upon specifically defined evaluation functions, which past efforts have not provided. The following discussion is proposed as the groundwork for the applied science of energy economics for building systems.

The building environmental system

A diagram of the building environmental system is shown in Fig. 24-1. The box to the extreme left represents the space to be occupied or conditioned. The space experiences a heat loss or gain depending upon numerous factors familiar to all practitioners. The rate of this heat loss or gain at selected maximum conditions is defined as the heating load or cooling load, respectively. The letter E shown at various points around the diagram represents points of energy flow into or out of the system. The arrow designations are used to denote the direction of energy flow. Note that arrow, E_1, is double headed, i.e., a point at which energy can flow in either direction is the load and in the context of this subject is the *block load* on the building. If the building consists of more than a single space, it is conceivable that heat will flow into the building system at one point and out at another; the net of these flows is the block load.

The terminal delivery system and terminal control system blocks are, in many systems, either interchanged in the relative locations shown on the diagram, or, in some cases, integrated into a single complex entity.

Thermal energy flows from the space into the terminal systems, or from the systems into the space, at a rate and quantity equal to E_1, thus maintaining the desired space conditions.

Energy required to motivate the terminal delivery systems is illustrated by E_2. The terminal control system is the point or points at which the conditioning of the air takes place, i.e., where the psychometric problem is solved. The energy which flows into and out of the terminal control system is made up of load energy (E_1) which can enter or leave; input thermal energy distributed from the high-level source system; leaving thermal energy distributed to the low-energy source system; and terminal distribution system motivating energy E_2.

The high-level source is made available to the system at a level above the space temperature and the low-level source system provides a sink at a level below the space temperature. From the standpoint of energy flows, it is quite common that within the terminal control system, energy will flow from the high-level distribution system ·into the low-level system, bypassing the space under usual operating situations. Thus, if the direction of E_1 is from the control system to the space (net heat load), the flow of energy from the high-level source may exceed the value of E_1, the excess

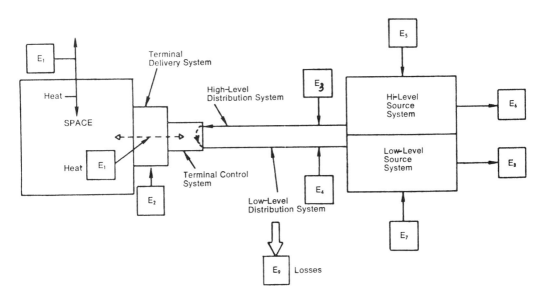

Fig. 24-1.

Table 24-1.   Summary of evaluation functions.

	Function	Symbol	Units	Description
Energy functions	Specific building cooling load	β_c	Btuh/sq ft	Building block cooling load/gross area
	Specific building heating load	β_h	Btuh/sq ft	Building block heating load/gross area
	Specific system cooling load	π_c	Btuh/sq ft	Cooling system design load/gross area
	Specific system heating load	π_h	Btuh/sq ft	Heating system design load/gross area
	Specific electric power (cooling)	κ_c	kW/ton	Cooling system electric demand/system design load
	Specific electric power (heating)	κ_h	kW/MBH	Heating system electric demand/system design load
	Specific thermal power (cooling)	τ_c	MBH/ton	Cooling system thermal demand/system design load
	Specific thermal power (heating)	τ_h	MBH/MBH	Heating system thermal demand/system design load
Power functions	Specific thermal energy	ϵ_t	MBtu/sq ft	Annual input thermal energy/gross area
	Specific electric energy	ϵ_e	kW-hr/sq ft	Annual input electrical energy/gross area
	Energy constant	C_e	Btu/kW-hr	Fuel input to generate usable electric energy
	Summary specific energy	Σ	MBtu/sq ft	$\epsilon_t + C_e(\epsilon_e)$

representing a quantity of energy flowing into the low-level system.

Energy flowing into or out of the distribution system may enter or leave at the terminal system or the source systems. Inputs E_3 and E_4 represent energy required to motivate the distribution systems.

The high-level source system converts available energy at the building to a useful thermal form. In the commonplace context of fossil fuel conversion, E_5 represents fuel input plus the energy to drive the feedwater system, the combustion apparatus (fuel pumps, F.D. fans, etc.); E_6 represents such items as stack, radiation, transformer, and friction losses.

The low-level source can be either a refrigeration system or a direct transfer mechanism to a temperature sink lower than the space. In the former, the heat removed from the lower temperature sink is that moving from the *terminal control system* into the low-level distribution system. One component of E_7 represents the external energy source to motivate the transfer, and E_8 represents the "output" energy from the refrigeration machine. Other contributing components of E_7 are the auxiliary refrigeration system loads: condenser water pumps, condenser fans, cooling tower fans, oil pumps, refrigerant pumps, control power, etc.

The remaining component, E_9 represents the irreversible losses which do not enter into the defined systems.

Power functions

One goal of energy economics is to provide evaluation functions for use by building systems design practitioners. The first of these functions is the *specific building load, β*. There are two Beta functions, β_h and β_c, representing specific building heating and specific building cooling load, respectively, in units of Btuh/sq ft. β_h is the design block net heat loss for the building and β_c is the design block cooling load for the building. These functions are calculated by determining accurately the block heating and cooling loads and dividing by the gross building area. The significance of this function is that it represents a major input into the ultimate building energy consumption and, although it may appear to be noncontrolled or independent input value to the environmental systems designer, it is definitely a controllable function as far as the building design team is concerned. The unit area is selected on the assumption that, beyond a minimum comfort height dimension, people use area, not volume, for habitation.

Those aspects of building design which effect the Beta functions include enclosure materials (walls and roofs); enclosure area per unit floor area; fenestration systems and areas; lighting levels, volume, occupant density, and ventilation rate per unit area; weather and climatic conditions; building orientation, program, and use schedules; and indoor design conditions.

It is evident that the essentially noncontrolled input variables, given a building use program and geographic location (or site), are weather and climate. All the rest are controlled inputs, established by the building design team. In this particular function the electrical designer can control the lighting level input and the architect the remainder, with the exception of the ventilation rate, which is under the control of the environmental systems designer. From the standpoint of energy economics, *ventilation rates must be established on the basis of contaminants*; e.g., general ventilation rates must be set on the basis of cfm per person rather than per unit area. For other contaminant types, such as cooking apparatus and systems, efforts must be sought to reduce exhaust volumes to an absolute minimum.

The Beta functions are power units—not energy units. They establish the *rate* at which thermal energy enters or leaves a space. Furthermore, the two functions are not additive, since the power and energy value of a unit of high-level energy is not equal to the value of a unit of low-level energy.

The specific system loads

The specific system loads, or Pi functions, like the Beta functions, are expressed in units of power per unit area (Btuh sq ft), and consist of two nonadditive components, π_c and π_h, representing cooling power and heating power, respectively. The Pi functions are defined as the maximum coincident heating and cooling demands (or loads) that the high- and low-level *source* systems will see. Design aspects which affect the Pi functions include Beta functions, performance parameters (control tolerance of temperature and relative humidity), terminal control systems, and distribution systems power to fluids.

The Pi functions are a measure of the maximum rate at which energy will flow into and out of the source systems from the distribution systems. To illustrate the relationship between Pi and Beta functions: If a building program consisted of one room with

one control zone, and the system were 100 percent effective the Pi and Beta functions would be equal. Although this is not the case with most actual building systems, in the practice of building energy economics, every effort should be made to minimize the difference between the two functions.

Insofar as the terminal control system is concerned, it will follow in most cases that the smaller the decrement between the Pi and Beta functions, the more efficient the system will be in reduced load energy consumption.

A comparison of this decrement is defined as the terminal system efficiency, in which

$$\eta_{\text{TC}} = \frac{\beta_c}{\pi_c} \times 100$$

$$\eta_{\text{TH}} = \frac{\beta_h}{\pi_h} \times 100.$$

Specific system electric power function

The Kappa functions (specific system electric power), and K_c and K_h are defined as the ratio of the electrical power input per unit of cooling power system capacity and per unit of heating capacity, respectively. K_c is expressed in kW/ton and K_h in kW/thousand Btuh (kW/MBH).

K_c is determined by adding the input electric power usually represented by E_2, E_4, and E_7, on Fig. 24-1. These include in addition to refrigeration drives, the auxiliary or parasitic loads listed below. (These auxiliary loads, additive to obtain K_c do not all have the same value when converted to the ultimate energy analysis. Thus, although they are all contributors, each must be considered separately by the systems designer.)

• *Supply and return fans* in commercial air conditioning systems have been found to range from less than 0.1 kW/ton to as high as 0.5 kW/ton. Of all the contributors to the Kappa function, the fan systems because of their high specific power requirement and extensive hours of use, should be the primary target for energy reduction. Two variables which affect the fan power requirement are the quantity of air circulated and the air system pressure. The quantity (cfm) should always be established at the minimum possible level to achieve acceptable performance. Efforts at minimizing the total system air flow must be accompanied by careful attention to effective air distribution methods.

Fan system pressure provides a broader area of control over the specific system power. The most common cause of high fan system pressure requirements has been the high-pressure or high-velocity distribution systems. There are two reasons for such systems: pressure needed for terminal unit control and restrictive space requirements. The designer can, in the application of energy economics, assign a quantitative value to each of these requirements. In most cases the value of the space saved by the high-velocity system is much less than the value of the energy used. Thus, from the standpoint of energy economics, high fan system pressures should be avoided where possible.

• *Chilled water pumping systems*—Like fan systems, the chilled water and heating water pumping system specific power requirement is a function of the flow rate and the pressure drop; also, like the fan systems, the water distribution systems are essentially continuous operation loads, thereby contributing significantly to the ultimate energy usage. Pressure losses in the systems have been fairly well established by the economics of piping systems and heat transfer surfaces. Thus the most readily controlled variable is the flow rate. Since the flow rate is a function of the system temperature range, it follows that, in the design of hydronic systems, the maximum possible temperature range should be taken at all times. In larger systems, the advantages of variable flow rates with load variations should also be considered. If load reductions are achieved by reductions in flow rather than temperature range, significant energy savings can be realized. An additional advantage of the longer temperature rises and the resulting decreased flow rates is that the investment cost of the system is reduced because of smaller pipe and pump sizes.

• *Condenser water pumping systems*—The first effort at optimizing energy use in the condenser water system is the selection of the most effective sink. This has a significant effect on the prime refrigeration energy required. Once this is done, the flow rate of the condenser water system is fairly well fixed by machinery availability and economic considerations. Like the chilled water system, the other variable contributing to the specific power is the pumping head. This head is created or established by the dynamic pressure drop through the condenser, in the piping system, and in system nozzles, and the static lift in the system. The most common cause of high condenser water system power requirements has been overcirculation, i.e., a miscalculation of the required head when selecting the pump, and an excessive power requirement due to the resulting mismatch of system curve versus pump curve.

• *Cooling tower fans*—Aside from performance and capacity requirements, the selection of the cooling tower should be made on the basis of space required, physical arrangement or configuration, construction material, cost, and specific power requirement. All of these parameters are interrelated, but product literature and resulting application have historically ignored the specific power requirement. A search of this literature reveals that towers with the lowest power requirement are often the least costly and constructed of the most desirable materials. Additionally, careful study of control logic schemes, although not reducing the specific power requirement, can significantly reduce the energy consumption of the cooling tower systems.

• *Air-cooled condenser fans* must be considered in the same manner as cooling towers. Generally the selection is more complex, since, with a given ambient temperature, efforts to minimize compressor horsepower will result in increased condenser fan power. Since the condenser fan system in many applications sees a full load throughout compressor operation, the energy analysis indicates a considerable contribution to the ultimate consumption by the specific electric power contributed thereby. As with the cooling tower, efforts in control logic should be aimed at reducing fan energy during periods of reduced load or below design ambient temperatures.

• *Control power*—Although the least significant contributor to specific electric load, the control power requirements cannot be altogether ignored by the systems designer. A simple guideline in the application of pneumatic systems is that continuous bleed-type controllers should be avoided wherever possible. As mentioned above, many currently available terminal devices employ fan system pressure for terminal control power. If this feature leads the designer to select a high-pressure fan system, the fan horsepower burden must be recognized as a control power contributor. *In this type of system, the specific power consumption can be most significant.*

• *Refrigeration system drives*—The refrigeration drives, if electric, are generally the largest single contributor to the specific system electric power function. Although design parameters other than energy economics may lead designers to the selection of refrigeration systems with excessively high specific power ratings, the penalties in energy and power must be recognized and justified.

The specific power requirement is not necessarily a linear function with annual energy consumption, but its computation will lead to areas of concentration for reducing the operating hours of the machinery or achieving more effective reduced load energy reduction.

The K_h function, the specific heating system electric power, like the K_c function, is determined by adding the high-level system auxiliaries and the primary electric energy input when the latter is used as the prime source of thermal energy. The auxiliary loads to be considered in the K_h function include fuel pumping drives, forced draft fans, induced draft fans, electric fuel heating, fuel compressors, condensate return pumps, feedwater pumps, and circulating water pumps.

These items are added together, then divided by the total system (high-energy-level output) capacity in mbh. In large systems, the use of increasingly less machinery volume to achieve given amounts of energy conversion or heat transfer tends to higher K_h values.

Because of the many advantages a variable temperature fluid offers, the most popular heating fluid systems are hydronic, requiring a pumping power linearly proportional to the product of the pumping head and the flow rate. The 20 F temperature drop from gravity systems used in product design and pipe sizing tables has been accepted as a norm. However, two significant advantages result from longer system temperature drops: less pumping power is required and smaller pipe sizes are called for.

Specific system thermal power

Specific system thermal power (the Tau functions τ_c and τ_h) is expressed in units of thermal power per ton of refrigeration (MBH/ton) or per unit thermal capacity output (MBH/MBH). The annual thermal energy requirement, though not linearly related to the Tau functions, will, in most cases, vary proportionally. The numerators of these functions are input values, so inefficiencies in thermal conversion systems, convection and radiation losses, and thermal system parasitics must all be considered.

Energy functions

The specific power functions discussed above can be computed quite readily as building system designs are developed and as the machinery and components are selected. Reducing each function will contribute greatly to the minimization of building power use. However, in the field of energy economics, efficient use of available resources is the prime target, and it does not necessarily follow that reducing the electric and thermal power requirements will minimize building energy consumption. The next step, therefore, involves the specific system energy evaluation functions for thermal and electric energy.

The quantitative value of all input energy has been established and reduced to specific units. When the designer begins the process of converting these units into energy functions, he must study such variables as hourly weather profiles, anticipated building use schedules, and the reduced load characteristics of each energy consumer or conversion device.

The initial loads, which resulted in the determination of the Beta and Pi functions, are usually calculated to satisfy performance requirements during anticipated extreme conditions. However, building requirements impose loads which are less than the design load most of the time.

Once the reduced load profile has been established, therefore, the various components and subsystems must be analyzed to determine their respective power consumptions at each reduced load condition. Virtually every energy conversion system has, on its capacity versus input energy curve, a point where the power input per unit of output is minimum: this is, the point of optimum efficiency. Even those subsystems which have relatively low input as a function of output will have an exponential rise with reduced load, except that the rate of increase will be reduced.

The part load profiles are then combined with the reduced thermal and electric load power inputs for the different components. The resulting annual thermal energy is expressed in Btu's divided by the building area, yielding the specific thermal energy consumption, ϵ_t, expressed in MBtu/sq ft. Similarly, the specific annual electric energy consumption, ϵ_e, is expressed in kW-hr/sq ft.

In units of energy, ϵ_e can be combined with ϵ_t by multiplying ϵ_e by the energy constant, C_e, required to generate the power. This constant will vary (depending upon available electric source, generating efficiency, distribution losses, etc.) from 8500 to 17,000 Btu/kW-hr. The product $\epsilon_e C_e$ can then be added to ϵ_t to give the summary specific building energy function, Σ, in MBtu/sq ft. In a sample study of ten similar use buildings in the same geographic area (Chapter 23) this function varied from 193 to 433 MBtu/sq ft.

Conclusion

The conscientious application of the concept of building energy economics will, in most cases, increase the engineering costs associated with building systems design, but should significantly reduce investment costs, power and energy consumption, and, consequently, energy costs.

Yet until an organized system of basic evaluation functions is universally employed, the application of energy economics in building systems will remain totally subjective. If these concepts were to be applied to the energy dynamics of a representative sample of buildings, then a control range of the functions could be made available to the industry.

25

Return to regionalism in building design

Shortly after the turn of the century, Henry Ford revolutionized the automobile industry when he developed the concept and methodology for manufacturing automobiles on a mass production basis. Students of business management recognize Ford's concept not only as the cornerstone of the automobile industry, but also as one that spurred the entire industrial-economic revolution of the twentieth century. The concept of producing large quantities of identical products is the bulwark of our industrial society.

When this concept is related to the building industry, some analogies can readily be drawn that relate to the departure from regional dictates. If the auto industry had been guided by the dictate to design and build different types of vehicles for, say, 20 different climatic areas of the country (not to mention the global market), it is likely that mass production cost advantages would have been much less conducive to two cars in every garage.

As technology in the building industry progressed, the trend followed those in the automotive and appliance industries; i.e., to develop singular concept products for the universal market. This was the beginning of the end of regionally oriented building design.

There has been some limited success in the area of prefabricated or industrially produced buildings. These, although the ultimate in the concept, are not the target of this chapter, however. We address ourselves here to the mass produced systems, subsystems, and concepts such as building curtain walls, massive fenestration skin systems, insulating materials, structural systems, structural components, modular concepts, and prepackaged mechanical systems.

The intent is not to criticize mass production efforts in the building industry. Without challenge, this approach has been the primary force in the United States building industry over the past quarter century. In retrospect, however, the ingenuity of industrial designers, production engineers, and marketing organizations, considered in light of the products that have resulted, would not have been possible had not energy resources been readily available at relatively low costs.

Mass produced, mass marketed systems are best identified by fads in construction materials, construction systems, and energy systems. Examples of such fads are:

- Totally fenestrated building skins.
- Modular interior design.
- High-velocity air distribution systems.
- Roof-mounted mechanical systems.

None of these concepts, or the resultant products, can be considered as inferior, but unfortunately, the concept of universality causes extensive misapplication of such mass produced systems. The misapplication, in turn, often results in an overreaction that causes the misapplied system to be replaced universally by another, and so the cycle continues.

The concept of universality in the building industry, essentially ignoring regional variations in climate and energy resources, must be reconsidered if we are to cope with the problem of limited, costly energy resources. The immediate evidence is that our industrial/technological community is heading off on continued fadism, i.e., developing and promoting new universal systems that will solve the problem. Unfortunately, this approach will fail, or succeed only at the expense of creating other problems that will open the way for still other universal solutions.

If we are to continue our leadership in building technology in an energy conscious and resource limited market, we must recognize the regional nature of building design. We must seek out the climatic areas in which certain building materials or energy source and conversion systems are most applicable and then market them for those areas only. This will be a whole new concept compared to our past method of doing business.

Consider all parameters

Designers must consider all the parameters of each building as a unique entity and overcome the temptation of repetitious design. For every building, the building design and energy systems design must be subjected to an iterative analysis until the design is energy and cost optimized. These studies must include not only the local climate but also the building orientation, shape, fenestration and wall systems, occupancy schedule, mechanical system and illumination systems, and owning costs. A major implication of this concept is that the building design team must take an even more responsible role, not only in functioning as an integrated team but also in the overall project conception.

This will not be easy. When building owners and developers tend to dictate to the design team certain emphasis on first costs, the designers must be prepared to inform their clients fully regarding the energy and maintenance cost impacts of these possibly unfounded dictates. As a result, all will benefit.

ASHRAE Standard 90-75 touches on the regional aspects of buildings in Section 4. This is a healthy beginning. But the regional nature of buildings and their unique occupancy functions and schedules must be recognized as an industry philosophy if the energy question is to be addressed successfully.

26
Infinite sink?

Chapter 19, entitled "Infinite Source" addressed the concepts of *infinite source* and *infinite sink* as they relate to energy economics. Following introductory discussion, that chapter concentrated on the source issue. This chapter will consider the receiving end of the energy flow path, the sink.

One interesting aspect of energy, in any form or forms, is that it has a tendency always to degrade itself to a lower "level." The earlier discussion discussed the sun as the only true infinite source, and even that source in its relative infinity is continually transferring its energy in astronomical quantities to the space surrounding it. A ball at the top of a hill will always roll down, converting some of its potential energy to velocity, which will ultimately degrade to a low form of heat.

Whenever this direction of energy flow is reversed—that is, when some force outside the natural order attempts to move the energy from a lower to a higher "level"—a quantity of energy is required that exceeds the increase. The option available, then, is to sacrifice a larger quantity of one form of energy to provide a smaller quantity of a more desired form.

Another fundamental concept that must be recognized is the fourth dimension aspect of energy: it is as time related as life itself. Starting with the sun, the energy emanates from the sun in the form of thermal radiation in dimensions of energy per unit time. Whenever we "consume" energy here on earth, supposedly for the benefit of mankind, we consume it in the power dimension—energy per unit time.

Prior to the advent of man on earth, the energy cycle achieved natural balances, which ultimately resulted in storing some energy within the earth. It was when man discovered the fossil energy stored in the earth that dependence on a finite source commenced.

Fossil energy is depleting

The time required for the formation of fossil energy resources is many orders of magnitude in excess of the time rate of consumption, to the extent that the former is irrelevant. It can thus be reasoned that fossil energy is totally depleting!

As energy degrades (considered as a source in its initial form), it can be thought of as simply taking a useful path toward a lower level region called a sink. And since man generates the need for the flow, it can be reasoned that man creates the sink. As an example, if man creates a need for energy in the form of electricity, he has dictated a need for energy in a higher form than the stored fossil resources. Thus, for each unit of electricity required, three to four units of fossil energy will move from the fossil storage form to ambient temperature form (the ambient temperature being the ultimate sink for earthbound processes).

Time-integrated loads are sinks

In the science of building technology, the time-integrated loads can be thought of as being intermediate energy sinks. The amount of energy available to serve these loads was relatively well matched to the true need when energy was not readily available and when the monetary cost of purchasing and converting it was relatively high. As the immediate availability became less of a problem, however, and the unit cost reduced, a mismatch of the consumption and the need started to develop.

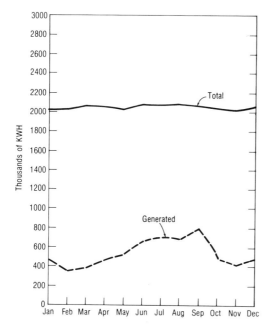

Fig. 26-1. Monthly consumption of electricity at educational institution.

Indeed, in many cases the true need was forgotten, and the problem of converting energy in ample quantities to serve *any* need became the underlying motivation. In engineering technology, efficiency of conversion became the goal, *not* efficiency in ultimate utilization.

Figure 26-1 is an example of the creation of an energy sink on the campus of a relatively large educational institution in the Midwest. This is a graph of monthly electrical energy consumption in kW-hr for a calendar year. It is most interesting to note that the consumption is essentially independent of those parameters that would normally be expected to affect it—outdoor temperature, hours of daylight, student population, etc. *On this campus was created the ideal electrical profile from the standpoint of the energy conversion system!*

Figure 26-2 is the fuel consumption graph for the same campus for the same year. Although there is a bit more shape to this curve than to that in Fig. 26-1, obviously relating to weather, the base is extremely high (minimum monthly consumption is 57 percent

of maximum) and is totally independent of campus population!

The staff responsible for the energy systems on the campus historically concentrated all their efforts on the efficiency of the conversion at the campus power plant. (The plant was an integrated energy plant that produced both electrical and thermal energy. The electric loads were shared with the electric utility company in such a way that the plant efficiency could further be optimized.) In these efforts, the staff was extremely successful; the plant was highly efficient.

In retrospect, however, the method utilized to accomplish the high degree of efficiency was one of building the loads to match the availability of a given form of energy. Though there may have been some validity to this concept in isolated instances, it was when it became the underlying philosophy upon which policies were based that reason lost control.

This case history is not unique. The concept of "building the load" is fundamental to our entire current-day economic system. Once a pipeline company invests in running a line across the nation, it is to its overwhelming

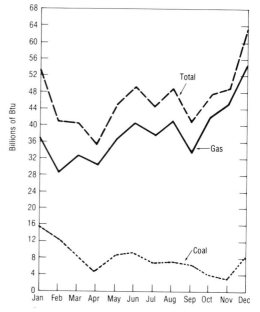

Fig. 26-2. Monthly consumption of fuels for same campus as in Fig. 26-1.

advantage to deliver all the fuel it possibly can through the pipe. Once an electric utility company invests millions of dollars in a generating plant, its economic survival depends on maximum utilization of that plant. If, as on the campus, the load does not exist in the natural order of things, the utility must create a market by building a load to support the investment in plant.

Thus it is that society continues without policy or plan to create an infinite sink into which we degrade our finite quantity of fossil energy reserves at an ever-increasing rate in a highly efficient manner!

27

Second law concepts

The second law of thermodynamics is a concept well understood and accepted by all mechanical engineering practitioners. The Kelvin–Planck statement of the second law tells us that *no system whose working fluid undergoes a cycle can receive heat from one source and produce work without rejecting heat to a lower temperature sink*. The interpretation is that if a thermodynamic engine is constructed to convert "heat energy" to shaft (mechanical) energy, some of the input energy must be "wasted" or rejected to a relatively low-temperature sink. The Clausius statement of the second law paraphrased states that *no refrigeration machine whose working fluid undergoes a cycle can receive heat from a low-temperature source and reject heat to a higher temperature receiver unless some external energy (from a level higher than the receiver temperature) is provided into the machine*.

Another well-understood concept by all mechanical engineers and other students of thermodynamics is the statement of the Carnot principle, which sets, in terms of the source and sink temperatures, the maximum effectiveness of either the heat engine or the refrigerating machine. For the heat engine, the efficiency (η) is defined as the shaft work output divided by the heat input from the high-temperature source. The maximum or Carnot efficiency, in turn, is:

$$\eta_c = (T_h - T_s)/T_h$$

where T_h is the absolute temperature of the higher temperature source and T_s is the absolute temperature of the lower temperature sink (see Fig. 27-1).

The effectiveness of the refrigerating machine is generally expressed as a coefficient of performance (COP), defined as the refrigeration effect (heat absorbed from the low-temperature source) divided by the external

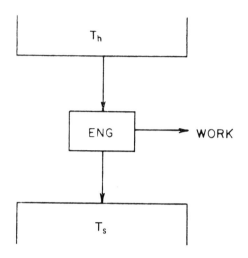

Fig. 27-1. Heat engine.

energy input. If T_s and T_l are the absolute temperatures of the higher temperature receiver and lower temperature source, respectively (Fig. 27-2), the Carnot, or maximum, coefficient of performance is:

$$COP_C = T_l/(T_s - T_l).$$

Fig. 27-2. Refrigerating machine.

For engineers, the major use of these Carnot values has been as guidelines in design development. If we want more shaft power per unit of input energy into the heat engine, it becomes immediately evident that we can achieve this by increasing the source temperature, reducing the sink temperature, or both. Similarly, with the refrigerating machine, to reduce the shaft or external energy input, we can raise the source temperature, lower the sink temperature, or both.

Another use is in evaluating the success of a design by comparing its effectiveness to the ultimate or Carnot value. Such a comparative analysis is developed in Chapter 60.

Students of thermodynamics have, however, identified some additional rather interesting aspects of the second law not commonly recognized by engineering practitioners. One of these is that a unit of thermal energy at a high-temperature level can readily be demonstrated to have more potential for heat at a lower level than the heat contained in the high-temperature substance. As an example, consider a flame, resulting from combustion of a fuel, at an average temperature of say 2000 F; it can be demonstrated through a simple second law analysis that many times more than the heating value of the fuel is theoretically available!

Referring to Fig. 27-3 there are three temperature levels, all expressed as absolute temperatures. T_h is the highest temperature source, T_s is a sink less than T_h, and T_l is a low-temperature source lower than T_s. Between T_h and T_s is an ideal or Carnot engine (ENG) that receives heat (Q_h) from the high-temperature source, rejects heat to the medium-temperature sink, and produces shaft work (W). Between T_s and T_l is a Carnot refrigerating machine (REF) that receives heat in the amount Q_l from the low-temperature source, motivated by the shaft work (W) produced by the engine, and rejects heat to the medium-temperature sink.

If the thermodynamic "system" is defined by the dashed boundary in Fig. 27-3, it is seen that there are two inputs, Q_h and Q_l, and two outputs. By combining the above two equations for Carnot efficiency and Carnot coefficient of performance, we can readily develop

an expression for the coefficient of performance of the refrigerating machine in terms of the three temperature levels:

$$\text{COP}_C = \frac{T_l}{T_h}\left(\frac{T_h - T_s}{T_s - T_l}\right).$$

This may be recognized as the common form of the ideal COP for absorption refrigeration machinery.

A first law balance on the "system" readily reveals that the rejected heat to T_s is identical to the sum of Q_h and Q_l. Since COP is described as the refrigeration effect divided by high-level energy input, then:

$$\text{COP} = Q_l / Q_h.$$

The heat to the medium-temperature sink, Q_s, is then:

$$Q_s = Q_h + Q_h \,(\text{COP})$$

or

$$Q_s = Q_h\left[1 + \frac{T_l}{T_h}\left(\frac{T_h - T_s}{T_s - T_l}\right)\right].$$

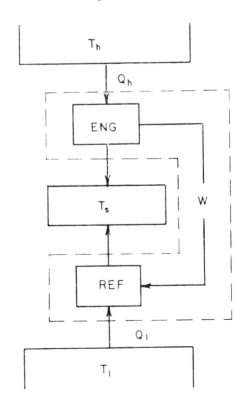

Fig. 27-3. Thermal cycle.

Cite common example

Consider a relatively common situation: T_h is a combustion flame at an average temperature of 2000 F; T_l is the outdoor ambient temperature of say 20 F; T_s is an indoor space temperature of say 70 F; and Q_h is the higher heating value of the fuel. Substituting the absolute temperature values into the above equation reveals:

$$\text{Heat to space} = \text{HHV} \times 8.53.$$

Thus, if the fuel is natural gas with a higher heating value of 1000 Btu per cu ft, the theoretical maximum amount of heat that the fuel could provide to the space under the conditions cited would be not 1000 Btu, but 8530 Btu—all with no wizardry or sleight of hand. This higher value could be referred to as the thermal availability. Further, if this fuel is burned at what would normally be considered 80 percent efficiency (first law basis), the 800 Btu obtained would represent only 9.4 percent of the energy theoretically available through second law conversion!

These concepts were generally developed in the nineteenth century but have been dusted off and reexamined in recent years. Because of technological problems relating to materials, economics, fluids, and the thermal loads themselves, it is recognized as totally impractical to even approach the second law availability value of the heat. But if the concept is kept off the shelf, it may help in more effective utilization of energy resources.

SECTION IV

Energy management

Energy management has always been one of the areas of responsibility of the manager of any enterprise or community in modern society. It has seldom, however, been identified as a subdiscipline of itself, but has rather been one small and complex component of the larger discipline of financial management. It is important to emphasize that in the vast majority of cases, it was both a small component and a complex component. It is for this fundamental reason that little or no attention has been given heretofore to energy management.

Consider first the relative magnitude of energy as a component of the total financial framework into which it fits, and how the relative size of that component is changing. The most fundamental example is the family unit. The major energy expenses for the family unit are household energy (heating, cooking, appliances) and fuel for the family car. The ratio between the average household mortgage and the energy cost in the early 1960s for the average middle class homeowning family was approximately five to one. In the late 1970s this ratio had dropped significantly and, in some instances, energy costs actually approached or exceeded mortgage payments. On the commercial and industrial scene the problem is quite different. Energy to the businessman has historically been a pass-through cost. The most direct example of this is the electric utility bill which contains a "fuel adjustment" which automatically floats through to the consumer, or the gas utility bill which similarly passes fuel costs through as "purchased" gas adjustments. Less directly, sophisticated commercial building owners have energy pass-through clauses in their leases and manufacturers in all industries including those which are "energy intensive" have historically taken the position that energy was a cost which their competitors bore as well as they, so a rising cost of energy simply raised the price of the "widget." In the case of the utility, the rapidly rising "purchased" energy pass-throughs have brought public outcries from the homeowner, and in the case of the commercial and industrial firms, the relative cost of the energy consumed by their building or their process as a part of the total space or product cost has risen to a point where management of energy consumption (or mismanagement) can seriously affect their position in the competitive marketplace.

It is thus that we find all levels of management from the largest industrialist to the homemaker starting to realize that an energy cost is not simply an overhead cost beyond their control. In working with otherwise skilled business entrepreneurs and management-level executives, it has been most surprising to find how many of them did not know how to interpret their utility bills. Yet this may have represented 20 to 25 percent of their total annual expenditures!

Unfortunately, because of the vastness of some of the topics in this section, only the surface of the all-important topic of energy management has been dealt with.

The first chapter in this section explores the broad scope of the need for energy management and the wide gap between present management skills in this area and the skills required. An invaluable management tool, the building automation system, is discussed in the second chapter. The building automation system is probably the least understood and misused device to have entered the so-called energy conservation market. The fact is that there is not a more powerful management aid that could have arrived on the scene more timely than the digital programmable computer which *arrived* during the past five years, just at the birth of our awareness of the energy revolution. Unfortunately, we have not yet taken the time to train our management people and technicians in the use of these devices. The top-level managers are expending funds well into six figures for an installation of highly sophisticated hardware with the misguided understanding that its presence on the premises will "save energy." Even more unfortunately, the manufacturers of this hardware are actively promoting this misunderstanding to strengthen the market. But just as most timely and new concepts in our marketing-oriented society tend to be oversold before they find their proper place in the scheme of things, building automation systems are ultimately here to stay. The key to their survival will be the teams of professionals and technicians with the systems knowledge to program the systems and the attuned management professionals educated to utilize this powerful tool.

The "Laundry List" is a timely chapter intended to inform the reader of some of the quantitative aspects of energy use which may be a good bit different than he may have thought.

The chapter on "Energy Audits" is probably the most directly useful data in this section insofar as benefit to the engineering analyst or business manager is concerned. The misuse of the concept of energy audits has found its way through United States federal legislation and the executive bureaucracy almost to the point where the concept has been destroyed. Hopefully, this is a temporary state of confusion. The energy audit, properly conducted and utilized, is as fundamental a tool to energy management as is a financial audit to financial management.

28
Energy management

Managers find themselves in an area in which they have little understanding, background, or training.

Approximately three years after the infamous oil embargo brought the topic of energy to the attention of the American public, we saw the following response:

• The United States economy shaken into a state of instability for which it has not yet found a formula for recovery.

• Federal and local governmental energy agencies created to cope with such problems as energy conservation, policy making, fact finding, and development of public awareness.

• Marketing emphasis of energy system devices and components of all types, concentrating on the energy conservation aspects of the product (heat recovery, efficiency, building automation systems, etc.).

• The embryonic development of a science of energy economics born, not of the economic disciplines, but from the areas of engineering and physics.

• Professional-technical societies, whose disciplines included those related to energy conversion, devoting a major portion of their resources, time, and funds to energy conservation measures.

A review of the activities during that time span clearly reveals that if, in fact, any progress had been made in recognizing and coping with the energy dilemma, it was by the engineering and physical science communities. It might also be recognized that these groups had for years, foreseen the pending problem but were totally unsuccessful in gaining the attention of the public or high-level decision makers in government, commerce, or industry.

The need, then, at this time is to project this progress beyond the engineering disciplines, where the fruits of the efforts will be the tangible accomplishment of the conservation of energy resources. The fact that this has not been achieved should be evident from the observation that all the efforts to date (January 1977) have not resulted in the reduction of national energy resource consumption.

Implementation is next step

The key to the next step is *implementation.* But the engineering community, in the vast majority of institutions in our society, is not in a position to provide this step—the knowledge and techniques must be turned over to managers.

Unfortunately, however, the pressures exerted on management are aimed at success in monetary economic terms. Heretofore, with energy generally representing a minor cost ingredient in most business ventures, the management community had little concern with managing energy. Now, as energy costs become more significant, many managers are recognizing the need for addressing it, but find themselves in an area in which they have little understanding, background, or training. As a result, energy management remains on the back burner.

Personal experience in the development of energy effective building systems for the past 15 years has revealed that the only way to accomplish energy conservation is through effective management. Yet this needed management is a very rare commodity. A growing recognition of this need has manifested itself in recent months by the increasing use of the term "energy management program" in place of "energy conservation program." It has been revealed that no energy conservation program can be effective unless it is successfully (and continually) managed.

Education needed

To achieve any degree of success, then, in energy conservation, the management community must be educated in energy management. Utilizing the popular approach of the economists—to simply let the price rise to the point that energy is a more commanding ingredient on the P&L sheet—would have little chance of success in the case of energy for two reasons:

• In most circumstances, energy is a pass-through cost. Competitive enterprises, all subjected to the same rising energy costs, simply increase the consumer's cost of the products (thus inflation).

• Management (in general) does not understand the energy sciences. Their training and experience historically have been in such areas as monetary economics and personnel (psychology).

Thus, the rising cost approach would result in either an inflationary spiral or business failures (or both).

A second approach, possibly more favored by some political leaders, would be to limit the use for a given task (rationing). This is a more forceful method of commanding management's attention, but considering the lack of management competence in energy disciplines, would lead inevitably to business failure or other institutional difficulties.

The third alternative is education of the management profession (see Chapter 11).

Although possibly an overly ambitious goal, this appears to be the only way in which the knowledge and accomplishments of the engineering professions in the energy conservation area can be brought to fruition. A game plan for this transfer of information and skill should include a concentrated interchange between top-management groups and engineering groups. The promising managers, like our early entrepreneurs, will recognize the need and develop the skills.

29
Building automation systems

The building design and management marketplace is being inundated by various grades and forms of what have been popularly labeled as building automation systems (BAS). Current generations of these systems virtually all employ as their central processing unit (CPU) digital computers, most of which have functional capacities far in excess of that required by the chores to which they are assigned. Probably the extreme is the case in which a digital computer is assigned the single chore of limiting electrical demand in accordance with a preselected priority input and providing an octal printed output that simply records which loads were trimmed and for what length of time. Such units sense only one element of the building performance, the electrical demand.

The manufacturers of these devices were quick to realize that the CPU had capabilities far beyond this one simple function. Sound marketing policies led to seeking out other uses for the sophisticated hardware. Designers of new projects and the owners and managers of existing buildings are being subjected to extensive marketing programs, the majority of which stress energy conservation as a primary advantage.

Misapplication of both limited capability systems, such as the one described above, and some much more sophisticated systems has apparently led many systems designers to the conclusion that BAS systems are not an effective tool in energy management. *The paramount cause of the failure is that the systems were installed with little or no understanding of the functional aspects and inte-grated nature of the systems being "controlled".*

Keys to proper application

Properly applied, virtually any building automation system can be a valuable energy management tool. Some of the key steps required for this "proper application" are:

• The BAS must not be applied in cases wherein the methods of sensing and transmitting performance data have not been validated regarding anticipated accuracy and repeatability when coupled to the specific model CPU.

• The BAS should be programmable. If the unit cannot be programmed with logic tailored to the specific system characteristics, then its use must be limited to those controlled systems whose logic is recognized in the hardware circuitry. With constraints of finance, energy, performance, machinery availability, etc., the additional false constraint imposed by a BAS program limitation would likely be counterproductive.

• The programming of the unit should be done by, or done in close cooperation with, the systems designer. This involvement on the part of the designer will require additional time (and consequently a higher design fee), but in most cases it will have a very valuable side benefit. When the designer sets out to develop a description of the system operating logic in the rigidly structured format of a computer program, any latent inconsistencies or flaws will certainly be revealed, and an improved design inevitably will result.

• Extreme care should be given to the selection of all types of sensing devices and signal transmission. This is the system element that both the unit manufacturers and the systems designers generally relegate to the realm of "nitty gritty." The problem, however, in most cases is that since the sensing devices are small and seemingly simple, efforts at cost control have misled manufacturers of many BAS systems to accept so-called "commercial quality" apparatus. More problems are encountered with analog sensing devices than with binary signal devices, but both require the same attention. As an example, the major area in which the BAS saves energy is that it takes all the slack out of the system operation. If, then, a chiller temperature set point deviates 2 F from that required due to sensing inaccuracy, the resulting deviation will quickly be felt within the conditioned spaces because all other elements of the integrated system are finely tuned. On the other hand, if the same deviation had occurred in a manually controlled system, certain elements of slack in the system operation would generally have been available to compensate. (Many times this was simply due to the logic contribution of the operator.) Binary signals used to sense such functions as air pressures for various logic and alarm purposes, if not almost *absolutely* reliable, provide false inputs that often result in excessive output errors significantly affecting the overall performance.

• Start-up procedures for both the controlled systems and the BAS must be rigorously planned and carried out. Trying to achieve that high degree of exactness in operating logic (an inherent purpose of the BAS) is meaningless unless the system being controlled actually performs in accordance with the logic programmed. Such things as inaccurate fluid flow rates (air and liquid), poorly calibrated controls, and the like become extremely significant when the system is controlled by a BAS.

• Management must recognize the need for a well-planned maintenance program for both the elements of the BAS itself and for the controlled systems. Some BAS systems include logic for scheduling maintenance, and the management staff assumes that this ends the problem. If, however, a properly skilled staff is not available to carry out the chores, what has been gained? Poor performance of systems components due to inadequate maintenance attention has the same effect as a system that has not been properly debugged. If the BAS itself is not properly maintained, the results are the same. In either case, it will fail to perform reliably, and *the personnel responsible for the building operation will lose confidence in its ability and return to manual operation and control procedures.* This has happened in a vast number of systems, whether controlled by a BAS or other, less sophisticated means.

30
The laundry list

As activities in the area of energy conservation in building systems unfolded from 1974 through 1979, we saw a slow but nonetheless steady trend toward a growing technical maturity. Undoubtedly, the oversimplified techniques proposed by those without a true understanding of the technical aspects of the problem, the fads used as a crutch by those who lack the time or the inclination to address the problem in specifics, and the cure-all products born of mass production manufacturing systems will always be present. A growing segment of both the engineering profession and the building management profession, however, is now recognizing the true value of proper direction toward the technical aspects of the problem and the relentless need for ongoing management of energy conversion systems.

Simple lists ineffective

In the early days of the then-popular movement toward "energy conservation" in building systems, the so-called laundry list approach prevailed. Such efforts were simple lists from less than a page in length to several hundred pages (the latter, of course, being government funded), which were intended to serve as checklists. The presumption was that if a building owner or systems designer simply went down the list and "did" things to his building or its energy system that were on the list, he would conserve energy. For several reasons, such lists very likely consumed more energy in the processing of the paper they were printed on than they have since conserved in building systems.

- In most cases, they did not prepare the user for the reduction in performance that would result from the actions recommended.

In some cases, the reduction would be direct; in others, it would be the result of a detrimental effect on another integrated system.

- In other cases, the result of taking the recommended action would be the consumption of more energy, often resulting in *increased* operating costs. For example, reducing the cooling load by adding reflective film on the glass simply might increase the heat consumption if the cooling system used reheat control.

- In their simplicity, they lacked the ability to incorporate cost effectiveness as one element of each item.

Thus, it might be concluded that although the laundry list approach might have some limited use, it is certainly not the effective tool it was once thought to be. We have now had the experience of some five to seven years of conducting detailed, accurate energy audits on numerous operating buildings of all ages and use types. The majority of these audits were performed by independent analysts on commercial and institutional buildings. Since they were funded by the owners of the buildings, the studies understandably were performed on larger buildings and on buildings that, in the opinion of their owners, consumed excessive energy. The value of this free enterprise effect cannot be faulted since it is these buildings in which efforts toward energy management can be most immediately effective in use reduction. The other obvious value of this effort is that these buildings turned out to be the institutional and commercial buildings designed and constructed between the mid-1950s and early 1970s. As a result, the buildings studied had employed in their designs the materials and systems techniques currently employed, and the results of the audits lead us to those areas in which we can most effectively concentrate our conservation efforts in the current design of new buildings.

Audits provide better lists

The audits, then, give us the background information on which we can now start to build more useful laundry lists. These lists, unlike the original ones, are not cure-all type checklists but rather what might be called *memory joggers* for competent designers and analysts.

The audit of a typical relatively large educational institution might yield the categories of use for fuel and electric energy shown in Table 30-1. With this information available, the areas in which efforts should be directed in either energy retrofit of existing buildings or design of new buildings and systems become evident.

A brief discussion of each of the use areas in the example audit of Table 30-1 follows. The categories for fuel energy are:

- *Space heat—60 percent.* This value actually breaks down into major components, transmission losses and ventilation, with values of 25 percent and 35 percent, respectively. The immediate indication is that reductions in either the amount of ventilation air *or* the hours of providing it can be as energy con-

serving as adding insulation, and at a fraction of the investment cost.

- *Control heat—30 percent.* Control heat is the major source of excessive energy use (or energy waste) that lacked recognition in most earlier efforts. Whenever there is simultaneous heating and cooling (whether it be to cope with moving loads on a design day or the daily reduction from the design day), there is a use of control heat. This also shows up in the compressor energy on the electrical side of the audit. It must be recognized that dehumidification requires this simultaneous heating and cooling, but the backlash approach of "outlawing" it should be avoided. Designers, however, should avoid the use of runaround energy whenever it is used for any purpose other than dehumidification if at all possible.

- *Domestic hot water—10 percent.* Although it might be thought that little can be done about the energy used for domestic hot water, an inventive designer might see numerous opportunities. As an example, in large commercial office buildings, some studies have shown that the thermal losses from storage and piping systems, time integrated, have accounted for approximately 90 percent of the energy inputs to the systems. In such cases, point-of-use heating, although more costly on the basis of unit energy, is found to be much more economical on an annual basis, and usually in investment cost as well!

Electric energy use areas

The use categories for electric energy are:

- *Supply, return, and exhaust fans—40 percent.* Of all the energy use divisions as they line up from highest to lowest use, the fan energy category most often comes up on top. This is a result of three past practices:

1) The extensive use of high-pressure fan systems during the past 20 years.
2) The lack of both designer consideration and operator understanding of the value of shutting down the fan systems during unoccupied periods.
3) The almost universal concept of constant air flow, with temperature differential reductions to cope with load reductions.

Table 30-1.   Energy use percentages for various categories based on an energy audit of a "typical" building.

Use category	Percent
Fuel energy—	
Space heat	60
Control heat	30
Domestic hot water	10
Total	100
Electric energy—	
Supply, return, and exhaust fans	40
Lighting	25
Refrigeration compressors	20
Cooling/heating auxiliaries	10
Miscellaneous	5
Total	100

Designers, as a result of this, should make every reasonable effort to minimize air delivery rates and air system pressures and to provide for unoccupied cycle operation with the fans turned off.

• *Lighting—25 percent.* Once thought to be the largest energy consumer, lighting has been fairly well relegated to second place. Still, it is a major area for attention in both retrofit and new building designs. The designer should accept the challenge of providing the most beneficial lighting levels for both general environment and tasks with the minimum number of installed watts and switching systems that may sacrifice a few hours of lamp life for the savings of many times that monetary value in electricity. (As an example, some designs were found to achieve perfectly adequate lighting levels in classrooms with approximately 2 watts per sq ft while others were found to use more than 6 watts per sq ft!)

• *Refrigeration compressor energy—20 percent.* It is not unusual that the refrigeration energy is only one-half of the fan energy. It is still generally quite significant in absolute value, however. One of the more interesting aspects of these systems is the interrelationships of the different use categories. As examples, the control energy in the fuel summary has an equal and opposite component in refrigeration energy; a large portion of lighting heat may conceivably impose loads on the refrigeration compressor; and the majority of the fan energy will ultimately impact the refrigeration system in most system configurations.

Considering the refrigeration system itself, much can be done to reduce the ratio of output refrigeration effect to input energy (COP).

Some of these concepts are discussed in other chapters.

• *Cooling/heating auxiliaries—10 percent.* Although this is a relatively small component of the whole, it is one in which considerable reductions can be achieved both in retrofit and new building design. In the simple selection of cooling towers, product literature reveals components available from less than 0.1 to 0.25 hp per ton of capacity. The largest annual energy consumers are pumps, the energy consumption of which is directly proportional to hours of use, flow rate, and head. Thus, reducing any of these three either at design conditions or with load reductions can prove beneficial.

• *Miscellaneous—5 percent.* This category includes such items as elevators, office machines, vending machines, and the like. For each and every one of these, the building owner essentially loses control over the energy consumption once the device is purchased. Therefore, the best approach to energy reduction in this category is the inclusion of energy requirements (time integrated) at the time of purchase.

The above is a "laundry list" of the major categories of energy consumption in large commercial and institutional buildings, based on a "typical" audit developed for existing buildings. It is intended to illustrate the major categories of energy consumption, the reasons for that consumption, and the extent to which both designers and building management teams have control over the consumption. The underlying caution, which relates to all such laundry lists, however, is that change should not be instigated or designs modified until and unless all of the implications of that action are thoroughly understood.

31
Energy audits

A necessary tool in any energy management program is the energy audit. An audit as defined in Webster's is "a formal or official examination and verification of accounts." In the case of an energy audit, what is being accounted for is the energy consumed. Such energy audits can be considered in many ways analogous to financial audits, both in their degree of thoroughness and in their purpose. Regarding thoroughness, just as with a financial audit, specificity is absolutely required. The extent of verification, however, is dependent on the accuracy desired; elements not well substantiated or verified must be so qualified.

Two types of energy audits

There are two major types of energy audits. One is the initial or exploratory audit; the other is the ongoing management audit. The exploratory audit is generally performed as a preliminary step prior to instituting an energy retrofit or an energy management program. It is the exploratory audit that reveals all additional efforts that can or should be made to reduce energy consumption. In this regard, a poorly performed audit is like an inferior foundation under a building. If the exploratory audit is not done well, the entire implementation or management program will almost certainly fail!

The ongoing management audit is analogous to a periodic financial audit of the records of a business. Without the periodic financial audit, the business manager cannot be certain of his financial position, successes, or failures. Similarly, without a periodic energy audit, one cannot ascertain the success of the energy management program. A major difference between the two audits is that the exploratory audit is necessarily conducted with the information available from past records whereas the ongoing management audits rely on metering and other recording devices incorporated into the systems to provide input to the energy management program.

Develop annual energy profile

The first step in the exploratory audit is to develop an annual building energy profile. Most buildings employ two types of energy, purchased and provided at the building boundary: electrical and thermal. The electric energy is generally used for lighting and power drives and the thermal for heating and, possibly, thermally motivated cooling (in some cases, of course, electric energy is utilized to provide the thermal needs). The electricity, in the vast majority of circumstances, is metered; and thus, the information contained in the metering records provides reliable historical data for the electrical energy profile. Further, this information is generally provided to the building owner monthly on his electric bill by the utility company. The information typically provided, in addition to the cost, includes:

- *Service period.* This defines the dates of service covered by the bill, or the dates on which the meter was read for the current and previous billings. The service period is most important in constructing an energy profile.
- *Energy consumption.* The energy consumed, in kilowatt-hours (KWH), is always shown; and on most forms the actual meter readings at the beginning and end of the period are shown.
- *Power or demand.* If the rate under which the energy is being sold contains a demand charge, the highest 15 minute integrated demand for the period is shown. Although the demand itself is not used to develop an energy

profile, it can be used for the demand analysis of an audit.

Although only the energy consumption and the dates of service are required to develop the energy profile, an energy management program cannot be instigated until the analyst has a thorough understanding of the utility's rate structure. Thus, at this stage of the study, he should obtain and thoroughly study the rate structure (see Chapter 32).

To develop the profile, the analyst obtains billing information for the immediately preceding calendar year, January through December. For each month (or closest period thereto), the energy consumed is divided by the number of days of the billing period, providing the average KWH per day of the period. *This normalization is essential in developing a profile.* The periods between meter readings will often vary by as much as 40 percent, thus having more effect on the energy consumption during that period than other variables such as weather, building occupancy, etc. The normalized energy consumption data are then plotted on the ordinate against the months on the abscissa.

If thermal energy is purchased in a metered form, as is the case with natural gas or steam, the profile development is similar to that for electricity. The amount of energy, either in the quantity billed or that quantity reduced to thermal units (Btu) is normalized by dividing by the days of the billing period and plotted against the months.

The annual energy profiles so developed reveal a significant amount of data about a building's energy systems and their management.

Figure 31-1 is a typical electrical energy profile for a building in the Midwest. It is immediately evident that this profile is for a building with electrically driven air conditioning and not with electric heat. The lower plateau from January through March and November through December is called the "base" use. This base use is generally not weather related but rather more directly related to use patterns, systems, and occupancy schedules.

Figure 31-2 is a typical thermal energy

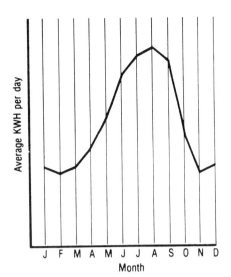

Fig. 31-1. Electrical energy profile.

profile for a building in the Midwest. This is a common profile for a gas heated building with some minimal gas use for service hot water and perhaps cooking. The curve shape relates very closely to heating degree-days.

From these two simple shapes, many other combinations appear, such as electric heating, reheat control, dual-stream control, absorption cooling, etc. With very little experience, the analyst finds that these profiles reveal untold information about the building's energy systems. The development of these pro-

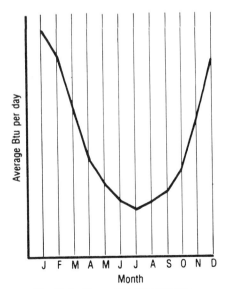

Fig. 31-2. Thermal energy profile.

files is quite simple and takes very little time. Yet the profiles represent one of the most important tools, both for performing the audit and effecting the ongoing management program.

Components of energy use

The next step in conducting an audit is to identify the component uses that created the historical profile. Again, since an audit by definition implies specificity, the degree of approximation and guesswork must be held to a practical minimum. If there is *any* difference between so-called mini-audits and maxi-audits, it should not lie in the general scope but simply in the degree of thoroughness and exactness in carrying out *this* step.

The general categories of use that are included in audits for commercial or institutional buildings, such as office buildings or schools, will be discussed here for illustrative purposes. Other types of buildings, such as health care facilities, manufacturing plants, etc., would usually have additional categories.

Conceptually, the categories are end-use categories. Generally, these are defined by the point where the energy in the form purchased is converted to accomplish an intended function, such as electric energy being converted to "fan" energy or "light" energy, or fuel energy being converted to a form usable for space heat. If the categories are subdivided into more components, the audit is not necessarily more accurate (since all elements must be included regardless), but it may be more useful. Thus, a major decision in the organization of the audit is the categorization of components. For example, the fuel energy that is converted in the boiler to hot water for building heat could be categorized as "heat" energy. A more useful option might be to subdivide this further into categories of space heat and ventilation. Another and still more useful step might subdivide the energy (in, say, a reheat or dual-stream system) into transmission net heat, ventilation heat, and control heat. The concept is that the more finely the pie is sliced, the more information will be available to make constructive use of the audit.

Categories grouped for audit

For illustrative purposes, a grouping of categories and a brief discussion of the methods of developing the audit quantities follows. Assume that for the example building, electricity is purchased to serve the lighting and power drive (mechanical) requirements, refrigeration is vapor compression, and utility natural gas is purchased to satisfy all end-use thermal needs.

• *Lighting*—All of the light fixtures in the building must be inventoried in a systematic and logical manner and categorized by switching circuits. Then, through thorough consultation with the building management group, the operators, and the occupants, a use schedule for each day of the week, with appropriate seasonal variations, is developed. The use schedule in some format relates the hours that the lights are on in a given period. If multiple switching circuits are combined, the schedule can use the technique of decimal portions of the whole. In any event, for each group of lighting, the energy consumed is simply the product of light (plus ballast) wattage and the hours of use, expressed in kilowatt-hours. As for all use categories, this is done on a monthly basis, then summed up for the year. The purpose of the monthly calculation is to permit comparison of the final totals with the historical profiles as discussed above.

• *Fans*—These could be subdivided into more than one subcategory, such as supply fans, return fans, exhaust fans, etc. The inventory technique and energy use development are somewhat similar to those for lighting. The first step is to inventory each fan, identifying its use and motor size. For better accuracy, the motor current and voltage can be measured with relatively simple and inexpensive instruments so that the actual operating power (kilowatts) can be determined from a motor characteristic curve. As with the lighting, the operating hours for each fan are then determined and the multiplication performed to obtain the fan energy consumed for the period.

• *Heating and cooling auxiliaries*—Again, at the analyst's option there are several possible categorizations. One such grouping (depending on the relevance to the specific system) might be chilled water pumps, condenser water pumps, cooling towers, heating water pumps, and boiler fans (forced or induced draft). Although the previous categories of lighting and fans account for the majority of energy consumption in most modern commercial and institutional buildings, the auxiliary category introduces the need for more complex calculations. These will be considered individually.

Pumps are of two types

Chilled water pumps and heating water pumps generally fall into one of two types. If they circulate a system with either wild flow loads or three-way valve loads, they will generally run continuously whenever the system is in operation. In this case, the pumps are inventoried in the same manner as the fans to determine their kilowatt power requirements, which are then multiplied by monthly hours of system operation to determine monthly energy consumption.

If the pumps circulate a system with throttling valve load control, the flow rate can be assumed to vary linearly with load; then by simply linearizing the part load performance of the pump-motor assembly, one can obtain the energy consumption from the product of the full load power requirement, the hours of operation, and the load factor for those hours.

In the calculations prior to this one, the energy determinations could be made by simply multiplying load sizes by their hours of operation. Their accuracy could not be improved by any other method, no matter how much more sophisticated than a pen, pad, and knowledge of basic arithmetic. The necessary introduction of the load factor multiplier at this point, however, could conceivably be accomplished with a great deal more precision by using appropriate computer techniques. It is important to realize that the use of the computer will simply improve the accuracy of the results. But in many cases, *skilled analysts* can achieve a perfectly acceptable degree of accuracy with manual calculations.

Condenser water pumps are generally if not always arranged for constant flow when operating. Thus, the monthly energy consumption is simply the measured kilowatt power requirement for the pump or pumps times the hours of operation. Since the pumps are interlocked with the refrigeration unit or units, their operating hours are the same as the refrigeration apparatus. Understanding the piping, interlock system, and operating procedures is important since it has been found that the majority of systems with multiple refrigeration machines utilize a common condenser water circuit. Thus, even though there may be, say, three chillers and three pumps, if it is a common circuit all the pumps may run even when only one chiller is required. It may be, however, that the operators manually valve out the condensers of the "off" chillers and thereby stage the condenser water pumps. Thus the necessity of understanding all three elements—piping configuration, interlock system, and method of operation!

Several methods can be used

The hours of operation can be determined by any of several methods. The least sophisticated and least accurate is a statistical estimate adjusted by the historical energy record for the unit; the most accurate is a thorough mathematical load analysis for the refrigeration system, one that reveals the need for specific amounts of cooling for a given number of hours each month or year. The latter is an extremely time consuming and complex calculation, usually not attempted without the use of a computer.

The energy consumption of the cooling tower fan(s), also a product of the actual kilowatt consumption times the hours of operation, is determined in the same way that the pump energy was calculated. The thorough or more analytical calculation approach, however, is a bit more complex because of the cycling of the fan, not only on low building loads but also when the wet bulb temperature is lower. Again, it is imperative that the auditor or analyst understand the manner in which the cooling tower fans are controlled and/or operated.

Other auxiliaries are related to the heating system and include burners, induced or forced draft fans, condensate pumps, and boiler room ventilating fans. Although these are usually a relatively small component of use, they should not be neglected. It is often the smaller motor loads that are found to run for many hours that are significant contributors to the energy consumption and offer the greatest opportunity for conservation. An understanding of the control methods for these auxiliaries will provide the key to calculating the hours of operation. The monthly and annual consumptions are then calculated in much the same manner as the cooling auxiliaries.

• *Refrigeration compressor energy*—In many buildings, the air conditioning refrigeration compressor(s) is the largest individual motor or power component. If the building is relatively simple in its configuration or the environmental system operates in a "heat-cool-off" mode, the refrigeration compressor energy can be determined within a reasonable degree of accuracy by a simple cooling degree-day calculation. In more complex systems, somewhat more complex calculations may be mandated. In such cases, the most rudimentary approach is to break the time-integrated cooling loads down into the components of: transmission and solar; internal; ventilation; and reheat or other false load.

In this manner, by dividing the operating hours into selected divisions of occupancy modes and weather-related "bins," one can make a reasonably accurate approximation. If a higher degree of accuracy is desired, a digital computer program that includes these parameters in all relevant details can be employed.

• *"Other" electrical energy*—This category, in a given building, could include components such as elevators, conveyors, door openers, office machines, computers, and any other loads connected to the electrical system. For each, the energy consumption is estimated by the simple (but sometimes tedious) process of multiplying the hours of use by the power requirement (kW or watts) and the use load factor (equal to or less than unity).

Thermal energy components

Thermal energy in a building like the example described has fewer categories to be considered than electrical energy. It could, however, be a much more significant cost item than electrical energy. The major components are space heat, domestic hot water, and "other."

• *Space heat*—The method of determining the thermal or fuel energy required for space heat is similar to that used for calculating refrigeration drive energy. If the building has a simple heat-cool-off system, with small internal loads, a simple degree-day calculation utilizing a fuel conversion efficiency based on a series of relatively simple combustion efficiency tests will provide a reasonably accurate answer. If, however, there are appreciable internal loads, and/or the environmental system consists of simultaneous heating and cooling for control, more rigorous calculations are mandated.

These calculations, like those for cooling, are best segmented into components of transmission, ventilation, and control heat and then performed considering operating hours, occupancy schedules, and weather bins. Also, like the refrigeration energy determination, manual calculations including the necessary algorithms can provide acceptably accurate results, but there is reason to rely on computer calculations if that methodology is available.

• *Domestic hot water*—Domestic hot water consumption in buildings such as the example building is usually quite small. The monthly and annual use can be determined statistically by using consumption rates from the National Plumbing Code or ASHRAE data and adding the piping system losses (which often exceed the consumption energy). Improved accuracy can be achieved if the hot water is metered, but this is seldom available for an exploratory audit.

• *"Other" heat*—Consumption of fuel or thermal energy for cooking, deicing, snow melting, or any other use should be investigated and added into this category. The important thing is that no use or consumption category be left unidentified. For example, in a large steam system it is quite common to vent the return system. This "venting" consumption must be recognized; it could represent a significant use.

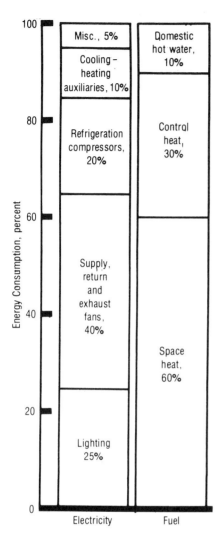

Fig. 31-3. Annual energy consumption in electric and thermal forms for example building.

The accompanying bar charts in Fig. 31-3 show the results of the use category phase of a typical audit of a building of the type considered.

Various uses of audits

The mathematically developed monthly consumption values are plotted on an energy profile diagram. The preferred method is to superimpose the calculated profile plot on the historical. This provides an easy method of determining the accuracy of the calculated values. In most cases, the initial attempt at this comparison will reveal a significant deviation between the two graphs. Generally (particu-

larly with electricity), it can be assumed that the metered values are correct (if proper care was exercised in developing the historical profile) and that any deviation is a result of inaccuracies in the calculated values. The nature of the deviations, however, will almost assuredly point to the source of the error. For example, if the metered profile for electrical energy indicates a considerably higher annual consumption but the monthly deviation is constant (same shape curve but higher), the analyst should look for an underestimate in some nonweather-related component such as lighting. It may be that in establishing the operating hours-percent usage matrix, the use by night custodial personnel was overlooked.

If considerably more fuel and electrical energy were used in the spring and fall and the system has electric refrigeration and fuel-fired reheat, the deviation would point to an error in estimating or calculating the reheat (or control energy) input from both sources.

Thus, after analyzing the extent and the nature of the deviations, the analyst recalculates the use components in subsequent "tries" until an acceptable match is obtained. Generally, a deviation on the order of 5 percent is the target although it is not uncommon to accomplish a match within 2 to 3 percent deviation. Once the match is deemed acceptable the mathematical model for the energy consumption can be considered valid.

Bar charts such as those illustrated in Fig. 31-3 are then developed for the final model. This could be considered the final step in the exploratory audit.

There are numerous uses for an audit, but the commonest are to serve as the basis for energy management feasibility considerations, as a format in setting future energy budgets, and as base consumption data against which future use patterns can be compared.

Referring to the bar charts in Fig. 31-3, the following use percentages are illustrated:

- *Electricity*—Lighting, 25 percent; supply, return, and exhaust fans, 40 percent; refrigeration compressors, 20 percent; cooling and heating auxiliaries, 10 percent; and miscellaneous, 5 percent.
- *Fuel*—Space heating and ventilation, 60

percent; control heat, 30 percent; and domestic hot water, 10 percent.

Study determines feasibility

A comparative study of the use percentages leads quickly into the feasibility considerations. For example, the fans are seen to represent the largest single component of electrical consumption. It could readily be determined how many hours per year the fans could be cycled off if an unoccupied control mode were added. The calculation would be redone with the new operating hours, and the results of that singular effort specifically identified. Similar reasoning is applied to each use component, the analysis done, and the results obtained. The new energy use quantities thus calculated for each feasibility option are extended to a monetary savings by application of the appropriate rates; and by comparison to the estimated implementation cost *for that respective option*, the feasibility is determined.

The value of this approach as the introductory effort in an energy management program cannot be overemphasized. It is the only approach that will provide a quantitative, confirmable, relatively accurate method for determining the value of investing capital for the purpose of reducing energy costs.

Consider other "action" steps

The subsequent "action" steps of the energy management program include retrofit, operation, and monitoring. Following the retrofit work and/or appropriate changes in operating techniques (for both the building use scheduling and the energy systems), the success of the efforts is determined through various techniques of monitoring. One such technique, requiring little effort and suggested as a minimum, is to plot the base consumption profile (historical) and superimpose it on the new anticipated profile obtained from the mathematical model containing the final retrofit and operation modifications. Then, each month the energy invoices are analyzed, normalized to the days in the billing period, and compared to the "budget" on the graph. If the budget is not being met (i.e., actual use exceeds the budget) the cause should be determined and appropriate changes or corrections made.

Management audit is next step

The next step of sophistication is to extend the monitoring into the format of the ongoing or management audit. This consists of a periodic (usually monthly) component use development analysis of the energy consumption. With proper planning, this can be accomplished quite simply. Several types of submetering are available that can be used to provide the data for a reasonably accurate monthly audit. Examples of such devices are volumetric displacement water meters in domestic hot water makeup lines and in steam condensate systems, lapsed time clocks on constant power electrical loads such as many fan and pump motors, and electric (kilowatt-hour) meters on variable electrical loads. Then, when the actual usage deviates from the budgeted quantities, these continuing or management audits will tend to pinpoint the sources of such deviations.

The energy audit and budget, as well as the entire concept of energy use management, represent a new and unproven field of endeavor. Although the need has been almost universally recognized, the implementation techniques have been misdirected, misused, and oversimplified to the point of counter-effectiveness in energy conservation. The techniques discussed herein are direct and have been effectively employed commercially for a number of years.

Earlier it was emphasized that these audits can be done with reasonable accuracy using manual calculation methods. Computerized techniques for the exploratory audit analyses, operations, and management auditing are strongly recommended when they are available, however. Considered in this way, both the value and the proper use of the computer as it applies to energy conservation efforts can be kept in perspective.

32

The structure of electric utility rates

In Chapter 31 the following statement was made: " . . . an energy management program cannot be instigated until the analyst has a thorough understanding of the utility's rate structure." This chapter will address the fundamentals of the philosophy and techniques most commonly employed in rate structures.

Basically, two components of cost are associated with the production and delivery of electricity. One is the fixed cost or investment cost in conversion and distribution systems. As has been stated in other chapters, this cost is related not to the energy delivered or sold but rather to the rate at which the plant *can* convert and deliver the energy. As is the case in any energy conversion system, then, the investment cost is seen to relate to the power. In electrical terms, the power unit is the watt, or more commonly the kilowatt (kW).

As a plant with a certain power capacity is operated, it delivers energy, in units of watt hours or kilowatt hours (kW-hr). The second component of cost, generally called the variable cost, relates to the energy converted. This cost includes fuel, supplies, labor, and other elements that change with plant utilization.

The most widely accepted philosophy of pricing electricity is to relate the charge to the customer to the cost to the supplier. A simple analysis of the cost to the supplier per unit of energy delivered reveals the relationship between cost and plant utilization. One way to express this utilization is by the term *load factor* (LF). The load factor, expressed for a time span of one year, is:

$$LF = \frac{kW\text{-}hr \text{ produced}}{kW \text{ capacity} \times hr \text{ in year}}$$

or

$$LF = kW\text{-}hr/kW \, (8760).$$

If, as an example, a plant can be constructed for \$500 per kW, and an amortization schedule reveals a need for \$120 per kW per year to amortize that investment, it is readily seen that if only 1 kW-hr is produced in a year, the investment component charge for that unit of energy is \$120. If, however, the plant produces 8760 kW-hr for that investment, the investment burden per kW-hr is 1.37¢. Thus, the fixed cost on a unit of energy basis varies with the load factor. Continuing the example, assume that the variable costs (fuel, labor, etc.) to produce a unit of energy are 1¢ per kW-hr. The cost per unit delivered can be expressed by the equation:

$$Cost = variable \, cost + \frac{amortized \, fixed \, cost}{LF \, (8760)}.$$

Solving this equation for our example, we obtain:

$$K_p = 1 + 1.37 \, LF \qquad ¢ \text{ per kW-hr.}$$

This equation is plotted in Fig. 32-1. The graph clearly illustrates the remarkable impact of investment cost on the cost of energy produced as plant utilization varies.

In developing rates that are the structure for charging the user, the philosophy of relating the charge to the production cost is usually implemented by treating each user as though a miniplant and distribution system were provided exclusively for him. Thus, a curve such as the one in Fig. 32-1 is theoretically applied to that customer's use pattern.

The simplest of rate structures is the straight declining energy charge, a rate generally available to small and statistically consistent users such as residences and small commercial establishments. This rate is based on a statistically defined minimum load factor; then, as usage increases (implying an increasing load factor), the unit rate drops.

A common rate for larger users, called an "hours of use" rate or a "hidden demand" rate, puts the burden of load factor control more directly on the purchaser. It is generally based on a monthly increment (approximately 720 hours) and has declining blocks of energy charges based on hours of use of the maximum metered 15-minute integrated demand. Using the data in Fig. 32-1 and assuming 10 percent above the production cost as the selling rate, the following rate might be established:

LF, %	Hours of use	¢/kW-hr
20	First 144	8.8
40	Next 144	4.5
60	Next 144	3.6
80	Next 144	3.0
100	Next 144	2.6

Generally, the last increment would be around the 80 percent step and would be stated as "all over 432 hours."

Another rate commonly applied to large commercial and industrial users is the rate that completely separates the demand and the energy charges. This type of rate uses a billing demand based on the highest 15-minute integrated demand for the month, which in essence passes the customer's share of the investment cost straight through. In the example discussed, a month's share of the annual $120 cost would be $10 per kW of demand. Some rates add to this concept a technique called a "ratchet," whereby the user's demand charge is based on the current month or the highest of any of the preceding 11 months

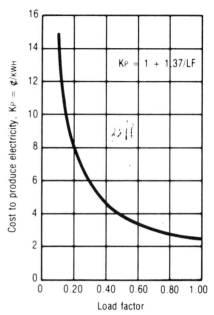

Fig. 32-1. Cost of production v. load factor.

whichever is greater. The theory is that the producer still had to have that plant capacity available. Some of these rates incorporate a reduced ratchet such as 75 percent of the previous months, but the concept is always to pass the fixed or investment costs through.

The energy charges in most demand/commodity rates are usually in decreasing cost increments, a technique to reward the user for improving the load factor (after all, it only seems fair!).

There are many other methods of relating the sale price to the production cost, some of them in use and some in the planning stages. Among them are:

• Power factor charges. The lower the power factor on an electrical system, the greater the current (amps) per unit of energy delivered. Since the capacity limitations of all the *electrical* components are related to current, low-power factor affects utilization of installed capacity. These charges are generally in the form of either power factor penalties or KVAR charges.

• Time of day adjustments. These are generally available only for large industrial users because of the cost and complexity of available metering. With construction and consequently investment costs soaring, however, a good deal of attention is being given to this technique for the future.

• Time of year adjustments. The philosophy is similar to that of the time of day adjustment. It is an attempt to encourage users to purchase their energy at a time when the plant has unused capacity available. It was such seasonal adjustments that prompted such concepts as "all-electric" rates in past years.

• Fuel adjustments. This is an amount that usually varies by the month and is related to the cost of resource energy to the supplier. It is an additional multiplier that is applied to the energy use (kW-hr) charge.

In summary, the concepts in production cost control of electricity are twofold. The first is to obtain maximum use of invested capital through optimizing load factor; the second is to provide the product at the lowest resource energy consumption consistent with available resources and technology. These two should not be confused. And the underlying philosophy of rate structures is to encourage the user to mold his use patterns to enable the supplier to accomplish these.

The different rate structures throughout the world are almost innumerable. Most, however, are based on some elements of the concepts discussed above. As stated at the outset, the energy analyst should obtain a copy of the rate filing, or a syllabus thereof, at the initiation of any investigation. Then, when he is successful at duplicating historical costs, he will understand the rate and can proceed.

SECTION V

Codes and standards

Codes and standards are quite different types of documents in their composition and purpose. In their content, however, they address similar topics, and oftentimes are based upon each other. Building codes, where they are of the totally prescriptive type, can have more effect upon environmental systems design and the energy consumed by the systems than any design concepts or systems selection of the designer. It is for this reason that the author is apologetic for not including more material on the subject of codes and standards. Philosophically, however, in justification for the lack of more subject matter addressing this extremely important topic, codes and standards are always based upon current state of the art at the time they are written. The first step, then, in the revision or updating of the code or standard is to advance the state of the art and that goal is the fundamental purpose of most of the chapters in this book.

Unfortunately, particularly with building codes, the very existence of the code tends to fix the state of the art and prevent or hamper progress because it is based upon yesterday's technological level. The cycle of change is several years, thus the industry moves as if on a stairway rather than on a continuous line upward.

Consider the differences between a standard and a code. The second chapter in this section dwells on various types of standards and their value to society as well as their technical-economic structure. Simply stated, a standard is a method, technique, or other expression of the state of the art which all *knowledgeable* practitioners agree is the most accurate expression or statement thereof. For example, if for the benefit of the consuming public, it is desired to establish a standard method for testing room air conditioning units, the method decided upon is agreed to by all of the persons who manufacture the units and that method becomes the "standard" method. If there is serious disagreement between the "experts" (those who design and manufacture the units), no consensus is achieved and no standard is adopted. Thus, in the entire area of consensus standards, it is germane to recognize that a standard method or state of the art exists *only* if there is consensus among the *experts*.

Once consensus is achieved, the standard becomes a public document against which all technical aspects relating to the topic of the standard are compared. Continuing the example, if there is a standard on the method of testing and rating room air conditioners, all reputable manufacturers will use the method and so stipulate in their technical or sales literature. As a secondary step, private testing laboratories or agencies are often used to serve as watchdogs to "certify" that the testing and rating was done properly in accordance with the standard. All licensed professional designers recognize the value of the standards, and virtually none would accept a component or subsystem that did not comply to the nationally recognized and accepted standards.

It must be recognized that in some areas, "meeting the standard" may not imply the "best," since the standard may have had to be reduced to some level of mediocrity in order to achieve the consensus.

Codes are a bit different than standards, particularly in the realm of building codes. Codes are not necessarily authored by technical experts, they need not achieve a consensus prior to finalization, and they are legally mandatory as statutes or ordinances. Additionally, there exists a large army of enforcement officials to ensure that the designers and constructors of buildings comply with the letter and, in some cases, the intent of the building codes.

Building codes have proved time and again to be a necessary part of the responsibility of local and/or state governments. Oftentimes where they do not exist (as in many rural areas of the United States), unsafe, unsanitary, or unhealthy building environments have been found to exist. Unfortunately, since the authoring and adoption of codes is not in the hands of knowledgeable professionals, these codes have often been manipulated in their basic content by special interest groups. This practice of manipulation, and opportunity has fortunately been reduced in recent years with the advent of the so-called universal, basic, or uniform building codes.

Another unfortunate aspect of building codes has been their tendency to fix the state of the art by statute to where it is (or was) at the time the code is adopted. Since statutes cannot be adopted that simply legislate, say, a code including future updating, it is possible that a national-type code document can be updated annually; but in a given community, if the local governing body does not act on a change to the legislation, the old version remains in effect.

As the energy revolution started becoming evident in the early to mid-1970s, many people recognized that the most evident vehicle to legislate energy-effective building design was through the building codes. The fundamental problem was that building codes, although they were used in virtually all major population centers, were locally written and controlled, and thus not suitable to influence by the central or federal government.

The major thread which the federal government has been able to grasp in this diffuse legislative situation is a national consensus standard adopted by the American Society of Heating, Refrigerating and Air-Conditioning Engineers (ASHRAE), entitled "Energy Conservation in New Building Design—ASHRAE Standard 90." Many so-called "model" code groups have essentially rewritten the text of this standard in code language and thus provided an energy conservation section in their codes. The challenge still remains for the federal government to convince the various state and local governments to update their building codes. The first chapter in this section addresses this issue. Also, Chapter 20 includes a discussion on the resource energy section of ASHRAE Standard 90.

33

Local building codes and energy conservation

Energy is one of the most talked about and written about subjects in the United States today. Preceding the infamous Arab oil embargo in the winter of 1973–1974, legislators, government, and business leaders considered energy as just one entry in the account ledger. The oil embargo was the catalyst for a whole new era in national and local legislative direction, and as on so many issues, the federal government moved first—logically, because the shortage of fuel appeared to be a national issue.

The energy dilemma is, indeed, a national issue. With the continuance of the present situation, the United States will be unable to defend itself in a major international confrontation and could be diminished to a secondary place on the world trade market with proper well-placed pressures from the oil-producing nations. It is in recognition of this that the federal government has moved quickly (and cumbersomely) into attempting to solve the energy problem.

Energy is regional commodity

The potential impact on local and regional commerce, unfortunately, has not been widely recognized by political leaders, although energy is a highly regional commodity. Historically, communities grew and prospered where there were raw materials, water, and other available modes of transportation. Where these three commodities were not available, there was nothing to sustain a community or community life. The amount of energy required was relatively small and generally readily available at nominal costs.

The growth of our cities and industries has, since the era of the Second World War, been based on the assumption of a continued availability of those depleting energy resources. This dependency moves energy availability into the cornerstone position of continued local prosperity (if not even survival). That is, the municipality that does not have the energy available will certainly not survive, and will yield its place in the marketplace and living environment to the community or region that has the energy resources!

If this proposition is accepted, then, the energy dilemma must be addressed as a local problem, deserving of solutions by local political bodies.

Energy supply limited

Since there is little that can be done by municipal agencies to increase the local supply or availability of energy reserves, the emphasis should be on gaining maximum benefit from those energy sources that *are* available. This is the concept of conservation.

Since energy was a readily available, relatively inexpensive commodity for many years, our society, including the business and engineering communities, did not consider energy economics in the design and production of energy-consuming products (a situation that was even encouraged by some energy industries). Furthermore, community planning in most cases was counterproductive to any efforts at energy economy. This resulted in the sprawling trend to suburbia that virtually negates any effective mass transit systems in our major metropolitan areas. Recognition of the areas of excessive use resulting from these past practices is the key to energy conservation.

The three major areas of energy consumption are buildings, transportation, and industrial processes. Each represents approximately one-third of the total national consumption. Again, to confirm the earlier observation that energy is a regional issue, the mix will vary with the region or municipality.

Include conservation in codes

All participants in municipal government recognize that buildings are probably the most universal component of municipalities (i.e., they all have buildings). Furthermore, most municipalities have building codes. It is thus in the area of buildings that municipalities can and should concentrate their initial efforts at energy conservation policies or legislation.

As many municipal governmental bodies have recognized, the concept of locally developed codes, ignoring the need for compatibility with other areas or a certain element of universality, has resulted in restrictive practices leading to adverse economic impact on the community as a whole. This recognition has resulted in most municipalities adopting nationally accepted building codes (such as BOCA or ICBO) drafted into ordinance with exceptions or supplements addressing local conditions.

Consider accepted standards

Thus, in the consideration of energy conservation ordinances or legislation, some nationally accepted guideline or standards should be considered.

ASHRAE Standard 90 "Energy Conservation in New Building Design," is a nationally recognized standard. It is available today and could be incorporated into an ordinance by municipal governments or state legislatures. If the building code vehicle is used, the responsibility in most states will fall to the municipalities, since the majority of states have opted to leave building code legislation and regulation to local or municipal governments. The standard could be adopted for use by the municipalities themselves in their own municipal buildings and could be extended into all private sector buildings in whatever method best serves the dictates of the region.

Many code administrators have expressed concern over additional work loads created if review for compliance were added to their already overwhelming work load. Well-designed legislation, however, could avoid this problem by such techniques as simple certification of compliance by the legally responsible building designer and filing with the code agency a record of anticipated annual consumption by the building.

If the municipalities do not recognize their responsibilities to address the energy dilemma in at least this minimal degree, the void will inevitably be filled by the state or the federal government. Such efforts on the part of the federal government are already well under way, and only quick action on the part of local governments will allow them to retain local control of the local components of this truly national problem.

34
The value of standards to society

Most people have taken very little time, if any, to reflect upon the general topic of standards. Yet standards are among the most vital and basic instruments of a modern technological society. Engineers will immediately identify with this concept, since they use standards of various types regularly in their daily activities. But before addressing those standards specifically relating to building systems technology, it may be of some interest to reflect upon more fundamental issues.

Without standards, civilized societies could not exist and function in the way we know them. Probably the most fundamental standards relating to commerce are the monetary standards. These, by necessity, are established by the national government of a society, and the affect of their establishment is to raise our commerce above the primitive level of bartering.

One step removed from the monetary standards are those governmental standards deemed necessary for effective communication within either the national or international community. In the United States, within the Department of Commerce is the Bureau of Standards. The Bureau's fundamental role is to establish the basic standards of communication. The most elementary of these, almost as necessary as the monetary standards, are the standards of weights and measures. Scientifically, these include many more units than weights and measures and are sometimes referred to as dimensional standards. Even the least technical of us relates to these standards, such as the length of an inch, foot, or mile, the weight of a pound, the heat sensation of 90 F, and so on. In the engineering and scientific community, practitioners become familiar with other less well-known units of measurement, relating to astronomical distances, radiation emission, energy, power, etc.

The vast majority of the standard units in the English (FPS) system used in the United States were initiated in a much less technical era and, in addition to being valid from the standpoint of standards, have the benefits of relating to physical experience—i.e., the foot was the length of the king's foot, the inch is approximately the distance between the first knuckle and the tip of the thumb, etc.—and of being oriented to ease of communication. As an example, if energy is being considered for heating a room, this energy is referred to as *Btu*; but if energy is being removed from a room to prevent overheating, it is referred to as *ton-hours* of cooling. Thus, when the unit *ton* is used, it not only implies a quantity of heat but also a process, and from the standpoint of thermodynamics, there is a significant difference between the addition of heat, usually a first law process, and the removal of heat, usually a second law process.

Metric units coming slowly

The comfort and convenience of such national standards has been found to be an ever-growing burden in this world of increasing international activity. To address this problem, following years of effort, a new system of dimensional standards, the Système International d'Unités, has been adopted, to which the United States, along with most other industrialized nations, is currently attempting to convert. Because of the total dependence of commerce upon dimensional standards, the conversions will be extremely difficult. As a result of its impact on institutions and individuals, it will be resisted. A change in the manner of thinking, to the man on the street, is only the tip of the iceberg. Relating to temperature in degrees Celsius instead of Fahrenheit, to distances in kilometers instead of miles, and to milk cartons in liters rather than quarts, will generate resistance that can be overcome with little difficulty. If nothing else time will solve the problem as new generations grow up with the new systems.

The engineering practitioners, manufacturers, mechanics, and others in technical disciplines will represent less volume of resistance, but more substance. The mechanic will need a second set of tools for a time, a problem that will be overcome with time. With the influx of foreign products in the past decade, many mechanics and homeowners have already solved this problem. The manufacturer will, in some cases, find it necessary to retool. This is an extremely costly burden that will ultimately be borne by the consumer, either in the product cost or as a federal tax. In many cases, United States industry will benefit, since much of the heavy industry in this country is currently outpriced on the world market by foreign manufacturers with more modern, efficient production machinery. Thus, as one steps back and observes the impact upon the mechanic and manufacturer, the inconvenience and adverse economic impact appear painful but temporary and definitely advantageous in the long run. Now, the mechanic, because of the presence of foreign products, already needs two sets of tools; whereas, following the total conversion to SI, he will need only one. Now, the manufacturer is competing, with archaic production machinery, in foreign markets that use other standards of measurement; whereas, following complete conversion to SI, he will be in a much more competitive position.

Thinking metric more complex

The impact on engineering practitioners and the physical relationships with which they work is much more complex. The simple problem of *thinking* in terms of meters rather than feet, watt seconds rather than British thermal units, and pascals rather than pounds per square inch is analogous to the problem of the man on the street, albeit more difficult. This will be simply a dimensional change in his thinking, which can be considered the first-order problem.

The second-order problem will be his relationship to *constants* and *evaluation functions*. All engineers in the HVAC disciplines relate to such constants as 1.08 and 500 in the heat capacity equations; most know the flow rate capacity of various sizes of pipe; some have the steam tables fairly well in mind. With the change of dimensional standards, not only will the education of change be required, but the efficiency of production will suffer significantly (a cost problem analogous to the manufacturers') and the more serious problem of increased error will be manifest. One hopes that society will accept this last problem and that the practitioners will not be destroyed as individuals or as an industry during the conversion period.

SI units are inadequate

The third-order problem for the engineering community relates to the use of the new dimensional standards in engineering technology. The new SI system is seriously lacking in the units it defines. For example, the system does not include units for fluid head (energy per unit weight) or heat removal (cooling) energy. This lack is in some cases one of simple adjustment and subsequent acceptance of the inconvenience of the communication mechanism, such as with the lack of an equivalent for tons of cooling. In other cases, however, such as with the absence of a fluid head term, adequate time must be allowed for the practitioners in the arena of their engineering societies to resolve the problem. If, indeed, a fallacy exists in the proposed system, it must be corrected. On the other hand, the purely scientific approach to the new system may prove to update the current concepts and practices of the related engineering disciplines. Before the new system is finalized, it will, it is hoped, be subjected to the mature, unemotional scrutiny of the engineering community, and they in turn should address the change objectively.

Consensus standards are helpful

Another type of standard, to which John Q. Public relates much less than to these first two, is the consensus standard. Although there is this lack of recognition on the part of the public, consensus standards have done more to help and protect the consumer than all the consumer advocates, consumer-oriented federal laws, and consumer protection bureaus combined.

Consensus standards are one of the best available examples of voluntary efforts working for the common good. These standards are adopted by organized groups such as trade associations and technical societies for the purpose of standardizing dimensions, testing methods, safety regulations, etc.

At first glance, it might appear that if a trade association adopts a consensus standard, it follows that that standard would be to the benefit of the members of the trade group but *not* in the best interest of the consumer. Upon further consideration, however, it becomes evident that the "good guys and bad guys" syndrome of the typical consumer advocate movement is not valid where consensus standards are concerned. Consider some typical classes of consensus standards.

• *Dimensional standards:* If the lumber industry had not adopted dimensional standards, the dimensions of lumber would vary. For example, a two-by-four might range in dimension from 1½ in. by 3½ in. to 2½ in. by 4½ in., with ill-defined tolerances. If this situation existed, each lumber dealer would have to try to effectively market his size two-by-four. We might have "Brand X is bigger . . .," or "Brand Y two-by-four takes up less room and is lighter. . . ." All of which would simply add to the cost of the two-by-four regardless of the dimensions! This situation would render the lumber industry unstable, to say the least.

From the consumer's standpoint, the increased cost from the lumber mill would be only the beginning of the problem. The retailer would have to stock several different sizes of two-by-four, further increasing the cost. The lack of standard dimensions would make construction extremely difficult, whether it be for new projects, remodeling, or repairing. Anyone who has ever done any retrofit on an old frame structure can identify with this problem.

The two-by-four is an overly simplistic example. Dimensional standards extend to all classes of materials used in machinery and construction, such as pipe dimensions, metal thicknesses, wire dimensions, bolt threads, and structural steel sections. They enable us to design and construct machinery, systems, and buildings at the least possible cost to the consumer, and most repairs can be done by the consumer with *standard parts* purchased from a hardware store.

• *Safety standards:* Safety standards apply in such diverse areas as manufacturing processes, materials (composition and fabrication methods), devices, and buildings. From the standpoint of the manufacturing interests, these standards assist in the design development of safe products and protect the reputable manufacturer from unfair competition from less scrupulous manufacturers who would market a less safe product at a lower price. It is self-evident that such safety standards serve to protect the consumer from unsafe products. Over the years, consensus safety standards have done more to protect the public than the Occupational Safety and Health Regulations (OSHA) ever will and at an immeasurably small fraction of the cost.

• *Standards for testing and rating:* In a bit more subtle manner, these consensus standards are among the most effective consumer protection devices in existence. Before the advent of the standards for testing and rating, the capacity of a machine or device was simply that claimed by the maker of that device. The purchaser had no assurance, other than the reputation of the manufacturer, that it would produce the capacity claimed. Although this may not seem like a significant protection to many consumers, except in the case of a few consumer products like residential air conditioners and furnaces, all of the components parts that go to make up such things as central air conditioning systems can only achieve the results intended if they *each* produce the capacity intended by the systems designer. Where standards are not available, the designer inevitably relies upon safety factors for insurance, which results in higher investment costs to the consumer and higher operating costs.

Consumers have effective input

In the process of adopting consensus standards, the consumer and others affected by the proposed standard have more effective input than they have in any other process, including the democratic lawmaking process. Most organizations that draft and adopt consensus standards function under the rigorous guidelines developed and monitored by the American National Standards Institute. To adopt a consensus standard requires a consensus, not a simple majority as in the democratic process of lawmaking; the text of the proposed standard must be made available to anyone who wishes to review it, and all such reviewers are eligible to submit comments. All constructive comments must be considered by the adopting agency, which must either incorporate them into the final document or show technical cause for their exclusion. The "write your Congressman" approach of simply being for or against with no substantial reason is not part of this procedure. Perhaps this is a singular reason why consensus standards have historically proved more beneficial to the consumer than so-called consumer protection laws! In the adoption of consensus standards, there are no lobbyist activities or voting blocs to cloud the true issues.

In a few cases, there has been evidence of misuse of the consensus process. Some reaction to this abuse has been activity on Capitol Hill, and more recently in the executive branch, to essentially destroy the voluntary consensus standards process through such diverse approaches as weakening it through withdrawing the participation of government employees or by replacing the entire process with standards mandated by law and drafted by federal bureaus. In either situation, the voluntary consensus process would collapse, to be replaced by either the purely political process or the uncontrollable bureaucratic process. Or both!

If, indeed, there is or has been misuse of the voluntary consensus process of adopting standards, these abuses can and should be eliminated by the adopting agencies themselves, if these standards are to survive and provide the service to both the manufacturer and consumer that they have in the past.

SECTION VI

Uses of the computer

Electronic digital computers and the advances in electronic technology devices are somewhat parallel developments, but quite interdependent, and could justifiably be labeled as the new sciences of the decades in the third quarter of the twentieth century. Analog computers are, have been, and will continue to be valuable tools in the control of machinery and solution of engineering problems. The analog computer, however, is a device designed and constructed to address the solution to a specific problem in which the analog circuitry represents the parameters and algorithms which relate to the problem. The digital computer is a good deal different in concept and application. It is like a totally controllable "brain" which lacks an intellect, and the resources of the brain can be put to work on a single given problem at a time. In the context of this discussion, the "computer" includes the central processing unit (CPU) along with the necessary assemblers, compilers, and input–output communication devices which enable us to establish mutual communication between man and machine.

The first chapter in this section addresses a brief chronology of the development of the use of the computer as a tool in the design of building environmental systems. This chapter was written back in 1975, and already the ". . . development of the science" recognized in the chapter is well underway.

The second chapter in this section (also published in 1975) addresses the technology of applying these fantastic machines to the analysis of building systems, how the analyst can determine if the program being used can be accepted as providing reliable data, and a few words relating to the technique of using the programs to accomplish a meaningful result.

At the present time, we are starting to observe a closing circle in the use of computers which will likely have an unprecedented effect upon the building industry, from both the standpoint of systems concepts and present institutional structures. Historically, the building design and construction industry has been a "small business" oriented industry. The reason is quite clear—large business is inherently a large-volume operation, and large volume can most effectively be accomplished by mass production. Anything other than mass production to accomplish large volume must rely on a much higher level of skills on the part of the organization's employees. We use a typical machinery manufacturer as an example, such as a manufacturer of automobiles or air conditioners. Adequate amounts of engineering talent are provided to design the product and its production line (this is actually a rather small percentage of the finished product cost); the product is then mass produced and sold to the consumer. Once the product is in the hands of the consumer, there is the need to perform preventive maintenance and occasional service or repairs on the machinery. The vast majority of these maintenance and services requirements are performed by small business or even individual private practitioners. This is true because, although the total cost of maintenance and repairs over the lifetime of the equipment may greatly exceed the original cost of the product, it requires a higher level of skill, than did the manufacture, for each dollar earned.

Successful efforts at mass producing buildings in any area other than very low cost homes ("mobile homes") have been quite limited. The reason is that buildings are probably the highest cost consumer product by several orders of magnitude, they have a long life span, and they must be adapted to a large variety of consumer needs including size, purpose, site, materials availability, energy availability, relationship to adjacent structures, availability of operations, maintenance and service skills, and many others. Thus, if a manufacturer wanted to start to mass produce office buildings he would have to develop a design or design series that would satisfy the combination of all of the requirements better than they are presently being accomplished. Since this is virtually impossible, the design of buildings has been and will likely continue to be provided by skilled individuals in small businesses including the professional firms in architecture, structural engineering, electrical engineering, and mechanical engineering.

This nucleus of professional designers had been developing a new skill during the decade of the seventies which even they did not realize. A combination of the existence of a few systems analysis computer programs and the energy situation has led the buildings engineering profession into the area of computerized systems analysis as a decision-making tool. Thus many decisions previously made on the basis of judgment and experience are now being made on the basis of judgment, experience, and considerable quantitative data furnished through the use of a computerized system analysis.

Although it has not been recognized in the industry, the system designer has for years been the person most knowledgeable in the technical aspects of the system he designed. The consulting engineering profession can be said to have dropped the ball in not recognizing this, and utilizing this knowledge in providing additional (and needed) services to the building owner such as testing, balancing, and start-up of systems, as well as the necessary guidance and assistance in operation, maintenance, and service. The systems designers of the future, skilled in computerized systems analysis, will have a much better understanding of the system dynamics because of the substantial amount of information they will have received from computer programs, and the additional systems studies they will make to avail themselves of the full advantages of the computer capability.

When the system designer thus becomes knowledgeable in digital computer technology it becomes evident that the next responsibility which he will have to pursue is that of programming the building automation systems (see Chapter 29); thus the closing of the circle. It is at this point that we will have taken full advantage of the capabilities of the digital computer in building automation systems, and the computer will have found a new use. Here, instead of receiving input answers from the analyst, and printing output information, the computer is receiving input analog data such as air temperatures, system water temperatures, and outdoor temperature change trends, and outputting electronic signals to operate valves, pumps, chillers, boilers, etc. This closed-circle concept is inevitably where building technology will be in the future. In the interim, the designers and analysts are using the digital computer to improve their decision-making ability and the manufacturers of digital hardware are marketing both hardware and software to automate the system operation. Because of the unique feature of each building system, mandating the need for a software program tailored to the features conceived by the designer, the present approach cannot succeed unless one of two changes occurs. The designer must either program the building automation system or he must design the system to fit the capabilities of the available BAS programs. The latter alternative would impose an artificial constraint upon the system design which is totally impractical—thus, the inevitable result is that the systems designer will be the agency to program the building automation system.

When the industry arrives at this point it will have reached its second plateau in utilizing the computer. Each building at this point will be controlled to optimize the interaction of the active and passive parameters of comfort, energy, power, maintenance, etc. Properly achieved, this degree of technical sophistication will have removed all "operational" energy waste from building operation. And the first plateau, that of use of the computer for systems selection will have removed the potential energy waste from the basic system design. Thus, of all of the so-called new technologies being explored to assist in the solution of the energy dilemma, the digital computer is already here, with the capability to affect enormous reductions in energy consumptions—it needs but to be properly applied and programmed.

The closing statements of the final chapter in this section touch upon data communications. It has been said that data communications is to the advance of computer technology what the railroad and truck (transportation) were to the industrial revolution. Data communications is the field of communication between computers, between computers and people, and/or between computers and machines. With effective low-cost data communications, every analyst and every building can have immediate access to the highest capacity digital computer.

The potential for the computer in building systems is beyond the scope of imagination. If and when we reach the second plateau, the closed circle between design and operations, perhaps we will be able to see the third plateau.

35

The potential for the computer in the design of building environmental systems

Energy and computers are probably the two most currently discussed topics in the HVAC field. Although building systems have been utilizing energy for centuries, it was not until the increased use in recent years resulting from product and system advances that did not consider energy economics as a design parameter, and the recent awareness of energy source limitations by the public that the energy parameter has come forward.

The digital computer, on the other hand, is a relatively recent development, which can handle, process, and store vast amounts of information in microseconds. It found an almost instant market in processing readily quantified variables, such as cost accounting, bookkeeping, etc. Properly programmed, it can "solve" complex problems in a small fraction of the time it would take to do so manually. In many cases, the time required for a manual calculation would be so prohibitive that an exact solution would be impossible.

Design engineers in the HVAC discipline were not quick to utilize these advantages for several reasons, thus the 15- to 20-year delay. Some reasons for this time gap are:

• Designers lacked training and skill to author or use programs.
• The need was not evident. "Successful" systems had been designed for years utilizing manual calculations.
• Fee structures discouraged improved de-

signs utilizing computer techniques at added design cost.
• Catalog data on components and subsystems in tabular or graphic form were accepted as accurate.
• Most design firms were relatively small and could not financially afford the necessary hardware.

Computer use now recognized

Numerous changes, however, have occurred in the past five to ten years that have created a vital interest in computerized techniques. They are:

• Recent engineering graduates were educated in computer programming and use techniques.
• Shared-time networks availed even the smallest firm of large-capacity computers for a relatively small terminal lease charge.
• First cost pressures accompanied by significant advances in systems concepts initiated an element of need.
• Energy awareness by the nonengineering public has opened a market for energy calculations (calculations so complex and extensive that manual solutions would be prohibitively expensive).

Enormous sums of capital were invested throughout the decade of the seventies in seeking methods to use the computer advantageously in this field; many efforts failed. However, at this time a plateau has been reached from which computer technology will doubtless move into a lasting and functional pattern.

Differences are cited

One lesson learned was that there are significant differences between engineering computer programs and so-called data processing. Data processing programs are based on a well-defined set of equations, numerical relationships, or algorithms, and can be effectively written by professional computer programmers. Most programs related to HVAC however, are so complex that the *most valid programs* are written *by skilled systems engineers*, for whom computer programming is a secondary skill. Programs thus developed are a bit more cumbersome but generally valid.

Another significant difference is twofold. First, hardware capacity limitations are most detrimental. And second, logistics are such that the scheduling of the computer for performing a calculation cannot be predetermined. When used as a design tool, computerized techniques are most cumbersome if access to the hardware is not readily available. The reason for this is that a good design program does not provide the *answer; it provides valid information based on the input.* Thus, when the output is obtained, it is often necessary to study it, and then change the input as dictated thereby until the output reflects the most desirable design. (This concept has been challenged by some, but the majority concur.)

Computer improves accuracy

Another lesson learned is that HVAC system design and system analysis have been less than exact sciences. For example:

• Load calculations formerly used provided an "adequate" answer. However, when the computer was employed to quantify many judgments and approximations of the manual techniques, it was found that considerable basic research in building load phenomena was required.

• Elementary energy calculation techniques previously used, such as the degree-day and simplified bin calculations, were quickly seen to relate poorly with actual energy usage in commercial and institutional buildings because of the consumption of auxiliary systems, control techniques, and ventilation and occupancy schedules. Again, considerable research efforts to identify the algorithms were dictated.

• Fluid systems designs, classically based on standardized constants applied to the Darcy Weisbach equation and Moody's friction factors, were found to be useful only for sizing piping systems on the basis of full design flow in all branches. Again, extensive research into the dynamics of the system under varying operating conditions was dictated.

In summary, the computer is not only coming into focus as an important design tool, but efforts at its application have opened many doors indicating a need for development of the science.

36

A reevaluation of computer use

(The original publication of this chapter follows Chapter 35 by approximately five years.)

Chapter 35 addressed the then-current status of the use of the computer in the design and analysis of building environmental systems as it was in October 1975. In relating that status to the immediate past and the immediate future, the following statement was made: "enormous sums of capital have been invested throughout the past decade in seeking methods to use the computer advantageously in this field; many efforts have failed. However, at this time a plateau has been reached from which computer technology will doubtless move into a lasting and functional pattern."

Computer technology misused

Unfortunately, in some respects we witnessed in late 1979 a gross misuse of computer technology, which must be recognized and understood and then be openly discussed among competent and experienced system designers.

The problem appears to be the result of a certain overzealousness or optimism on the part of some scientists and technicians as to the capabilities of systems analysis programs in and of themselves and the exactness of a unique algorithmic approach, negating the possibility of correctness in all others. The result is, simply stated, that those who have unfortunately been so misdirected have essentially replaced the concept of the credentials of the engineer or analyst with the credentials of the computer program!

In some cases, this overzealousness has resulted in the abandonment of valuable published data, because they were considered inexact when compared to the belief of what the computer could provide.

Computer not always necessary

Consider some examples of the problems identified. First is the problem of estimating annual cooling energy requirements. This is admittedly less than an exact science, and needless to say, the best that one can hope to achieve with manual techniques is a good "estimate" based on two things—statistical correlation and the skills of the analyst. At a time when the industry has a greater need than ever before (because of the energy situation) to estimate the annual energy burden resulting from cooling systems, the *ASHRAE Handbook & Product Directory* (1976 Systems volume, Chapter 43) has removed the information that appeared in prior editions and in its place inserted a discussion of what could be done if a valid computer program were developed or available. Yet with a good statistical data bank, including such variables as design load, weather, solar characteristics, and system types, a *good analyst* could very likely make a reasonable estimate. Thus, instead of abandoning the manual methods in favor of a possibly unavailable computer program, we should be improving the manual methods. There are times and circumstances when the use of the computer is totally unnecessary and unwarranted.

A second example of a serious misunderstanding of the use of the computer has come from the developers of some of the most heavily funded energy analysis programs being promoted for use today. The programs have been said to be valid only for comparative evaluations of alternative buildings, systems, or energy sources, *not* to provide absolute quantitative answers. Yet, any analyst who has attempted to use a systems analysis (or energy analysis) program for the purpose of conducting an exploratory energy audit is quick to recognize that for a given set of input conditions, the answer he seeks must be in the absolute, not the comparative (see Section IV). This "comparative" concept appears to be a holdover from the time when such programs were used or marketed primarily for the purpose of selecting a building energy source rather than for the purpose of analyzing all aspects of the building energy dynamics. It might be mentioned at this point that the difference in the purposeful use of a program may dictate its contents; when used for auditing or as a design aid, a program that is all-inclusive (loads, energy, economics) is considerably more cumbersome than one that allows the analyst to move in and out as the intermediate data are generated.

The third example is probably the most serious insofar as the future stability of the industry is concerned. This specifically addresses the problem of replacing the credentials of the engineer or analyst with the credentials of the computer program. This concept, which has been legislated in at least one state and appears to be actively promoted by some bureaus of the federal government, is that of "certifying" computer programs that an analyst can use to analyze the energy consumption of building systems. Such a concept must inevitably lead to determining what programs will be certified, which in turn leads to the rejection of the use of those programs that do not comply with the requirements for certification. Consider the implication of this philosophy. It is analogous to commissioning a skilled and experienced engineer in bridge design to design a bridge but making it a condition of the commission that he can only use a given computer program (with which he may be unfamiliar or uncomfortable) to do all his structural analysis. One might ask, if the constraint on the commission is the computer program, what is the need for the designer? It would seem that the only logical way to achieve a design whose responsibility can be laid to an individual is to allow that individual to select the analysis procedure upon which *he will stake his reputation!* Any other concept invalidates the age-old approach that has been applied worldwide to responsibility in the engineering profession.

Recognize misuse of computers

Not only has this misguided concept been slipped into the legislation under pressure of unwitting legislators, but there has also been considerable pressure on the American Society of Heating, Refrigerating and Air-Conditioning Engineers to get into the business of either certifying computer programs or developing standards for such computer programs. That Society has undertaken numerous studies in committee, and several of its members have on their own conducted studies subsequently reported in symposium papers and other Society programs and publications—most of which have proved conclusively that *in comparing system analysis programs, the analyst has more impact on the answers (or output) than the specific differences among the programs.*

The areas of misuse of computers or of computer technology discussed above—whether it be the problem of overreliance upon computer technology, when the mode of manual calculation is required, or the trend to replace the analyst's credentials with those of the computer program—must be recognized and considered in view of their impact upon the profession. If not, the public, so long served by responsible practitioners in engineering, will be the loser.

37

The computer as a tool for energy analysis

The engineer is the technician; the computer is the tool. The engineer provides the judgment and skill, and the computer yields the optimum in quantitative data. Together they can lick many of today's complex energy problems.

Computers have proven very useful for many applications. In the HVAC field, they have greatly increased the capability of the engineer in designing systems. They have also provided him with the means to study systems in existing buildings to effect substantial savings in energy and operating costs.

The intent of this chapter is to discuss relating to the computer as a tool. The engineer is the technician; the computer is a tool. How does the computer fit into the problem solving process?

A computer performs the arithmetic necessary to yield reliable data required for problem solving. An engineer recognizes that many assumptions and approximations are required in solving numerous problems. As a field of engineering matures, it becomes increasingly exacting, requiring a diminishing use of the "approximation" approach.

The computer is the tool that has permitted engineers to reach the nth degree of exactness for many types of problems because it can perform many complex calculations in seconds. Without it, comparable computations would require unreasonable lengths of time. Simulating the performance of a HVAC system to determine annual energy usage could take several months, even on the simplest of buildings. Yet, the result would not be as

accurate as a properly designed computer program could provide in a few moments. Even changing one or two variables to examine their effect on consumption could necessitate redoing many or all of the calculations. In the past, this situation prompted many assumptions and approximations to be used in determining a building's energy consumption. To conserve valuable time, this was a necessary alternative. A computer provides the quantitative data that otherwise would have to be approximated; *that is all that it does!*

When a computer is applied to the area of energy economics, two considerations are necessary. First, the applied program must contain the proper algorithms relating to the problem and all mathematical relationships that affect energy consumption in a building. Second, the output information should not simply be taken as the answer. This was a common error in some initial efforts, but as computerized engineering problem solving developed, it was realized that *the* answer, or final solution, *could only be obtained by applying skilled engineering judgment.* A computer simply gives good quantitative data that can be used to arrive at better solutions.

The end-to-end computer program, or one in which input data is supplied and a final solution comes out, is generally not a good engineering program. Valuable data generated between these two points, which might indicate that a change in input is desirable, could easily be overlooked. Input should be applied with judgment and experience. Thus, interface programs are more effective in engineering work.

Determining energy flow rate

Physics is a science of exact definitions, and engineering is a subscience of physics. It may be beneficial to review some of the basic concepts common to both. Energy is defined as work or the capacity to do work. The basic unit of energy is foot-pounds. The unit used in the thermal form is the Btu, which is by definition, the equivalent of 778.26 ft-lb (approximately the amount of heat required to raise the temperature of 1 lb of water 1 F at standard atmospheric pressure).

Power is the time rate of energy. Some commonly used power units are Btu per hour, tons of refrigeration, horsepower, and kilowatts.

In energy economic studies, it is imperative that the difference between energy and power is clearly understood and kept in mind. This is significant because power units are generally input functions to an energy program. Clearly, control (minimization) of the power aspects of any system reduces energy use if other factors, such as operating hours, are

kept constant. (To attain this condition, a computer is not needed.) If 1 Btuh is consumed for 8760 hr a year, the energy consumed is 8760 Btu. The significance of this simple statement is that if power is reduced at the front end of the program, the output quantity is also reduced. Explore this relationship by looking at a simple diagram.

Figure 37-1 is a basic diagram of energy flows in a building system. The largest block on the left represents the space (building, zone, room, etc.) being studied. The rate at which energy enters or leaves this space at any given time is the load—a power function. It is a cooling load if it enters and a heating load if it leaves.

Once a building's envelope internal energy systems, and operating schedule are fixed, this energy flow rate becomes the absolute integrated energy requirement. In this analysis, the absolute integrated energy requirement is considered to be the block load, not the sum of the peaks. This function is the starting point in a study of energy economics. Energy requirements at the building ultimately lead to consumption of resource energy—coal, gas, oil, etc.—in a magnified form due to losses and inefficiencies during distribution and conversion processes.

Load study is first point

The first point of application for a computer when trying to understand and thus minimize

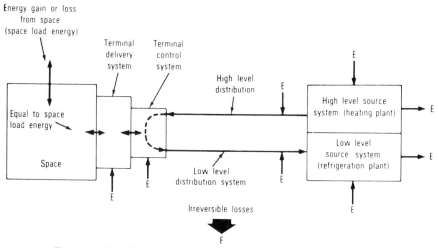

Fig. 37-1. Simplified diagram of energy flows in a building system.

usage is the load study. Load energy is a function of:

• building design, location, and orientation;
• ventilation rates based on contaminants;
• occupancy schedules;
• weather conditions.

Of these four, occupancy schedules and weather are essentially uncontrollable inputs. The controllable variables of the load are:

1) Fenestration systems, including:

• area of the glass;
• type of the glass;
• framing systems;
• interior shading;
• exterior shading.

2) Lighting systems and appliances.
3) Roofing systems and materials.
4) Wall systems and materials.
5) Excess ventilation rates due to:

• infiltration;
• overdesign of mechanical system ventilation rates;
• antiquated building code requirements.

A properly designed computerized load program enables a designer to modify any of the controlled variables to determine the effect on design load. When considering existing buildings, the location, orientation, and occupancy are, obviously, already established. Hence, the engineer must confine himself to the five controllable load variables listed.

The effects of changes in these five variables, either individually or in any combination, on both the block load and the sum of the peaks can be readily determined by a computer during a load study. Without a computer, these detailed load studies would be approximations at best, in addition to being extremely cumbersome and time consuming.

At this point in an energy economic study it is important to note two items. First, the time-integrated block loads—heating and cooling—on a building are the useful product energy requirements. Second, this product energy aspect—input to the mechanical system—is, except for excess ventilation, strictly an architecturally controlled parameter.

System component definitions

Before proceeding further, it is necessary to define the terms used in the following discussion.

• The *terminal delivery system* conveys air to and from the space. It includes the fans, supply and return ducts, grilles, etc.
• The *terminal control system* solves the space psychrometric problems. It contains everything needed to meet the conditioned space needs: cooling and heating coils with associated controls; control and mixing dampers; air terminal units, such as terminal reheat, mixing boxes, VAV terminals, etc.
• The *high-level distribution system* conveys a thermal fluid from a source to heat transfer surface (such as hot water or steam from a boiler to a coil). This process is self-contained in an electric resistance heating coil. Heat is generated by the current flow, and combustion occurs remotely at the generating station.
• The *low-level distribution system* conveys a fluid from a refrigeration machine (the source) to a chilled water coil or a direct expansion coil (the heat transfer surface).
• *Motivating energy* for the terminal control system includes the air energy required to operate the air terminal units. These require a minimum static pressure to assure proper operation. For example, self-contained variable air volume terminals may require a minimum of 2 in. WG static pressure. This must be added to the static pressure required to operate the duct system before an adequate amount of energy is provided to operate the terminal unit.

Building versus system energy

Referring again to Fig. 37-1, the second block from the left represents the terminal delivery system. All load energy to and from the space flows through it. A first law analysis* of the space clearly shows that the energy flow between the space and the terminal delivery system is equal to the *block* load energy.

Further analysis reveals that load energy flows into or out of the delivery system, while system motivating energy flows into the system. When a net block heating load exists, the delivery system motivating energy flows to the space. Energy analysis reveals that the motivating energy for the terminal delivery system often represents a major contribution to a building's annual energy consumption.

The terminal control system is represented by the smallest box from the left. Here, the space psychrometric problem is solved. Portions of this subsystem could be integrated with the delivery system in actual practice. In this analysis, however, it is important to identify the terminal control system as a separate subsystem.

First law analysis of the terminal control system reveals one of the more interesting aspects of energy consumption in HVAC systems. The energy flow arrows in Fig. 37-1 represent input from the high-level distribution system, motivating energy, output to the low-level distribution system, and either input or output between the terminal control and terminal delivery systems. The terms high and low level refer to temperatures above and below space ambient conditions, respectively; i.e., heating and cooling distribution systems. The high-level distribution system generates the "runaround" energy in the overall system. When the net building load is in the cooling mode, this system can receive:

- space energy;
- terminal delivery system motivating energy;
- high-level distribution system energy;
- terminal control system motivating energy.

All of these will flow to the low-level

distribution system. This runaround energy from the high- to the low-level distribution system is, as the space sees it, wasted energy. This condition often exists at full load as well as part load. As far as the terminal control system is concerned, this runaround energy may be essential to maintain the space conditions.

At full load, many building systems utilize runaround energy to make up the difference between the block load and the sum of the peaks. The high-level system false loads the low-level system to prevent overcooling the space. Figure 37-2 demonstrates this. It shows a single floor of a commercial building and net design cooling load conditions. The block load is 36.75 tons. This is the rate of heat gain into the space at design conditions. The sum of the peak zone loads, however, is 48.75 tons. This is approximately 33 percent higher than the block load.

Many systems are psychrometrically designed to handle the load represented by the sum of the peaks. Many others are designed to provide a cooling plant capacity somewhere between the sum of the peaks and the block loads. The energy difference between the two loads comes from the high-level distribution system. Thus, the high-level system false loads the low-level one. From the standpoint of energy economics, this false loading becomes even more significant. It requires added heat energy and also added refrigeration energy to

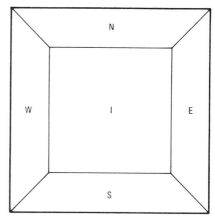

Fig. 37-2. Thus: sum of peaks is 33% greater than block 2 A single floor of a commerical building at net design cooling load conditions.

*The first law of thermodynamics states that (classical definition) work and heat are mutually convertible. This definition can be broadened to include all forms of energy; i.e., one form of energy may be converted to another.

remove the extra heat energy, hence the term runaround energy evolved.

Typical examples of false loading are: hot decks of dual stream systems; reheat coils; perimeter radiation in conjunction with cold deck VAV systems; and single fan, multiple zone economizer systems (those that provide winter cooling with low temperature outdoor air in lieu of refrigeration). As previously mentioned, the external energy for the terminal control system often is a significant contributor to energy consumption when high-pressure delivery air is required to "control" the terminal units.

Energy is transferred between the terminal control system and the high- and low-level source systems through their respective distribution systems. (In an actual system configuration, these systems are often integrated with the terminal delivery systems.) External energy is often required to motivate this energy transfer, and significant savings often can be achieved by careful system analysis. For example with water, or other single-phase heat transfer fluids, doubling the temperature range halves the flow rate, which in turn, reduces the power (pump horsepower and energy) required.

In existing buildings, this halving approach would reduce the distribution system energy to as low as one-eighth of its original value.

High- versus low-level sources

The first law energy balance also applies to the high-level source system where either fossil fuels or electricity are converted (minus losses) to the heat energy supplied to the high-level distribution system.

The low-level source system follows the second law of thermodynamics: heat will not flow of its own accord from a cold to a warm sink. Thus, it requires external energy to transfer the energy from that system to an available heat sink. This external energy may be in the form of a prime mover or, with absorption refrigeration, heat energy. It is important to remember that the external energy applied to power the refrigeration unit is not the only input. All parasitic loads on the system, such as condenser water pumps, con-

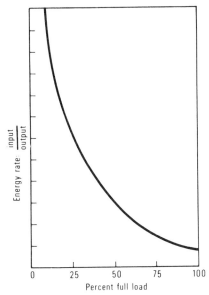

Fig. 37-3. An exponential curve depicting the decline in efficiency as the load decreases.

denser fans, cooling tower fans, control power, ventilation fans, etc., must also be considered.

Building environmental control systems operate at less than design load for a major portion of the year. Therefore, any valid energy analysis should recognize part load performance characteristics of the high- and low-level source systems. Virtually all energy conversion systems become less efficient (useful power output/power input) as the load decreases from the design capacity. The exponential curve in Fig. 37-3 depicts this common system characteristic. The fixed parasitic loads establish the rate of increase of input to output as the load decreases, and the curve

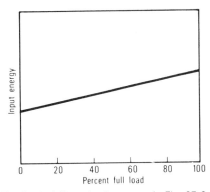

Fig. 37-4. A linearized curve as in Fig. 37-3.

approaches infinite input per unit output at zero load. An example of this is a boiler maintained at operating temperature when there is no heating demand. The curve in Fig. 37-3 can be linearized (Fig. 37-4) for computer programming purposes by multiplying the ordinate by the abscissa.

From load to energy analysis

It is with recognition of the proper algorithms and concepts previously discussed that a computer program to analyze energy consumption in a building is employed. Also, human interface with the output of the load program and the input to the energy program is essential in order to arrive at the best energy reduction solution.

Figure 37-5 illustrates the minimum requirements and output data of an energy analysis. (This discussion is concerned with the application of these programs rather than an in-depth discussion of the programs themselves.) It was stated at the beginning of this chapter that the computer is a tool. As such, the user applies it to determine what system modifications and/or changes in operating techniques are practical to achieve the desired goals.

Existing buildings provide the opportunity for immediate reductions in energy use compared to projects in the design stage. The procedure for applying computer analysis to existing buildings is as follows:

1) Study the building plans (if available).

2) Make a building survey to check the actual building conditions against the plans and/or to gather what pertinent data are required to perform the study.

3) Run a load program calculation.

4) Input the energy program with the output data from the load program and all the necessary data relating to the mechanical/electrical systems and the building use parameters.

5) Compare the output from the energy program with historical energy consumption records of the building, such as monthly and annual fuel and electric consumption from the utility bills.

6) Identify the reasons for any differences between calculated quantities and the historical data. Examples of differences are:

- improper calibration of controls;
- damper leakage;
- operating techniques;

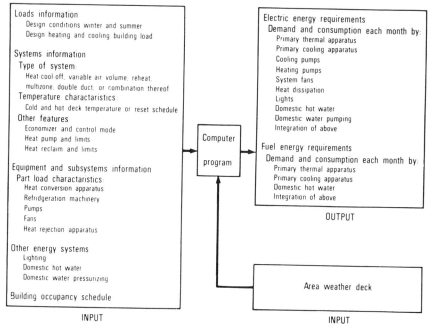

Fig. 37-5. The minimum requirements and output data for an energy analysis program.

- installation errors;
- design errors.

7) With the knowledge of the relevant energy-consuming contributors, starting with the building design, explore practical building and system modifications and control and or operating techniques by processing them through the load and energy programs. From the energy savings developed, determine the dollar savings yielded by each modification.

8) Determine the costs of implementing each of the various conservation steps analyzed.

9) Conduct a quantitative comparison between the energy cost savings and the implementation costs.

10) Determine the combined owning and operating costs and the rate of return for each conservation measure.

11) From all of this information, prepare a feasible course of action in implementing the steps that prove economically viable.

It is the quantitative aspect that is provided by the computer. Reductions in energy consumption can be accomplished in virtually all existing building systems. However, when modifications are undertaken, capital investment is often required, and only a thorough quantitative analysis will determine the wisdom of the investment.

With a valid energy program, the calculated results can be assumed correct, adjusting, of course, for statistical versus actual weather conditions. Obviously, this also depends on all input factors being correct. Many times, factors causing high-energy use problems can be readily identified and corrected with little or no capital investment, and engineering costs represent the only expense involved.

Proper understanding of the factors affecting energy usage in buildings and of the relevant algorithms in computing this energy use will permit the user to intelligently apply the computer to solving most, if not all, of the problems he is likely to encounter.

38

Computer applications for systems design and analysis

Introduction

Although computer technology has been available and recognized for about two and a half decades, the use of this technology in the field of building systems design has not gained the broad acceptance which was once anticipated. The reasons for this lack of acceptance appear to be:

1) The building systems industry is a loosely knit highly competitive team effort, and as a result, the large corporations who might have the financial resources to develop programs are not generally in the position of the final decision maker. This position is usually filled by the smaller corporate entities, i.e., the professional consultants.

2) The systems selection and analysis techniques used have been and still are in a dynamic stage of change. The industry has grown so rapidly in the years following World War II that the major design parameter has been to achieve comfort performance with peripheral efforts at reductions in first costs. Thus, not only had many other significant parameters not been identified, but even the fundamental one of performance had not been

This chapter was reprinted from *ASHRAE Transactions*, Vol. 82, Part II, 1976 and was originally entitled, "Computer Applications for Systems Design and Analysis." Appearance of this material in *Energy Engineering and Management for Building Systems* does not necessarily suggest or signify endorsement by the American Society of Heating, Refrigerating or Air-Conditioning Engineers, Inc.

well quantified by recognition of the relevant algorithms.

3) The practitioners in the design profession had not been educated in the use of computer techology and, as a result, were slow in the process of reeducation, particularly in view of the fact that they could continue to achieve "successful" designs (based upon the sole "performance" parameter) by continuing to do business as usual.

4. Computer hardware development prior to the past half decade appeared to concentrate on the basic computer console, with generation after generation of machinery aimed at increasing the speed of calculation and printout and increasing the capacity. Little development in data communications took place prior to 1975.

Significant changes have taken place in the recent past which have affected the status quo. In addition to the fact that many peripheral efforts have been expended at seeking the best method and direction for utilizing computers in building systems design by such groups as APEC (Automated Procedures for Engineering Consultants), the major equipment manufacturers, the energy industries, and some independent computer software companies, other catalysts have surfaced.

For engineering computer programs, unlike many data processing requirements, limited capacity computers which might have been in the financial reach of the potential users were not an acceptable tool. Many of these programs are enormously complex and require extensive iterations resulting in a need for high-capacity high-speed hardware. Another unique requirement of engineering programs is the need for first person access to the computer; remote batch processing is not acceptable. The reason is quite obvious upon reflection—that the printout of any good engineering program provides not the *answer* but rather documented reliable information upon which a decision can be based. Thus, upon receiving this information, the designer may elect to modify the input data and subsequently determine the effect upon the output.

Recent technological advances in data communications hardware and attendant networks of shared-time computer availability have provided this much needed instant access to high-capacity hardware for even the smallest of design offices at a nominal cost.

Maturity adds two catalysts

Maturity in the industry has provided two additional catalysts. One being the recognition of design parameters other than performance. These were manifest by such items as increasing costs of energy, increasing costs of investment monies, and an awareness of reliability requirements and maintenance/service costs. The other being increasing industry interest in developing quantitative approaches to systems analysis. As an example of how these two phenomena interrelate, consider the fundamental problem of cooling load calculations. In prior years, the only purpose for calculating the load was to establish the capacity of the machinery required to satisfy that load. The load thus calculated by manual steady-state methods was the maximum load that the building would impose upon the machinery at "design conditions." If the load thus calculated was a bit on the high side, the only effect upon the performance was, very likely, excessive use of process energy.

New engineering talent entering the profession was thoroughly attuned to computer technology. As such, the design offices found themselves doing a complete reversal. That is, the younger engineers questioned not the need for utilizing computer techniques, but rather why they were not being employed.

Consider current uses, trends

Against this background, consider the current uses of computer programs in a moderate-sized consultants office, and then look into the future at coming trends.

In addition to personnel who feel "comfortable" with computer communication and program understanding, in-house computers of adequate capacity are quite costly and frequently require a skilled professional to oversee the operation of the computer, creating the need for another person on the staff. A small firm, however, simply cannot economically justify this move. Thus, the present staff can be given access to the highest capacity hardware available on the shared-time networks by a modest investment in a communicating terminal similar to a teletype machine. To further minimize the investment, these terminal units can be leased on a month-by-month payment basis. The time-sharing companies are becoming increasingly responsive to the needs of both the users who really desire nothing more than how to run available programs as well as those who have programming capabilities and prefer supplementing the existing programs with their own efforts. It may be appropriate to reflect, at this point, that duplicate efforts at program authoring can be substantially reduced if the "program-capable" users have access to the logic of existing and available programs. The essence of engineering is understanding, and many engineers hesitate using a program without understanding at least the logic. This need has been recognized and the current trend is to provide the users with the logic information required. Figure 38-1 summarizes some of the advantages of time sharing.

Once the design office has obtained a terminal, access to any of a number of shared-time networks is available. Programs currently available from these networks include the following:

• *Heating and Cooling Load Programs.* Virtually all major networks have at least one loads program available to their customers. As both energy usage and capital investment become increasingly important design param-

```
LOW INITIAL COST
   EQUIPMENT
   PERSONNEL
EXPANDABLE
RELIABILITY AND REDUNDANCY
MULTIPLE ACCESS
NO OBSOLESCENCE
FLEXIBILITY
   TIME SHARING FOR SPEED
   BATCH FOR COST, LARGE CAPABILITY
MULTIPLE LANGUAGE CAPABILITY
EDITOR CAPABILITY
SUPPORT FROM COMPUTER SERVICE COMPANY
   SPECIALIZED PERSONNEL
   BROAD RANGE OF LIBRARY PROGRAMS
CALCULATOR MODE—HARD COPY
COMMUNICATIONS LINK
   WITH OTHER OFFICES
   WITH CUSTOMERS
```

Fig. 38-1. Advantages of Time-sharing Computer Usage

eters, computerized load analyses will be required for a designer to remain competitive. Many such load programs provide not only the capability of performing a room-by-room or zone-by-zone full load calculation (analogous to but more accurate than the manual calculation) but also the design or maximum "block" or "diversified" load and the building load profile data. Such load programs, then, provide as their output data bank the interface information for system analysis programs.

In the use of the load programs, one should always keep in mind the significance of the output information. The designer should scrutinizingly check each element of input data and confirm the reasonableness of the output information. The drudgery, however, is done unerringly and rapidly by the computer. If the output data differs radically from the designers estimate or past experience, he should revalidate the input and question the program logic until he is satisfied with the information received.

Although so-called dynamic load calculations are still undergoing a phase of embryonic growth, some such programs are currently available. These programs perform calculations which would be much too time con-

suming to do manually and take into account such phenomena as transient heat flux effected by building use programs, solar exposure, temperature fluctuations, building materials, the resulting thermal storage dynamics, and net-time-integrated heat exchange. The ultimate goal of these programs is, of course, to more correctly identify the design load requirement of the mechanical systems and provide more reliable input data to the interfaced systems analysis programs.

• *Programs for Use in Both the Design and Systems Analysis of Fluid Systems.* These programs are available from some time-share companies. For some time such programs have been available to assist designers in the sizing of air distribution systems and liquid-piping networks. These programs assist the designer in the proper sizing of the systems (eliminating first cost excesses inherent in manual less tedious analyses) and help assure properly sized or matched motivation systems such as fans and pumps.

A new series of programs have been developed which perform a dynamic response analysis to indicate system deficiencies under reduced load control response. One such program was developed at the University of

Illinois under an ASHRAE-sponsored research grant and is available through the Society.

• *System Analysis Programs.* Systems analysis programs provide the designer with information regarding the relationships between mechanical systems components and subsystems and the integrated building loads. These programs are one of the most immediately beneficial results of computer technology applied to building systems. The systems analysis programs provide such information as part or reduced load schedules for the mechanical systems. This information can be utilized to optimize reliability and minimize both first cost and energy costs all through the proper selection of primary module sizes. For example, the part load analysis may reveal that the selection of three chillers (the summary capacity of which equals the maximum system load; and the incremental capacity of which favorably matches the part load profiles) will achieve almost as high a degree of statistical reliability as two chillers (one a redundant standby), each of which has an incremental capacity equal to the maximum system load, while at the same time significantly reducing both the investment cost and the energy burdens.

The systems analysis programs also have the capability of considering the effects of first cost, reliability and energy consumption caused by such things as alternative psychrometric control schemes, variable-flow fluid systems versus constant flow, and various approaches at coping with system auxiliaries. Figures 38-2 through 38-4 demonstrate the use of a system analysis program in evaluating the impact of a proposed retrofit of variable-volume double-duct mixing boxes in a community college library building. Figure 38-2 shows the flow-temperature characteristic of the proposed mixing box. Figure 38-3 compares the refrigeration and heating before and after the change, as calculated by a system analysis program. Figure 38-4 compares calculated energy costs before and after the proposed change.

• *Other Programs.* Other available programs include those which offer the designer total solutions to complex textbook equations and graphs. Although this type of program has just scratched the surface of its potential, one called ROARK is currently available which contains formulas from the book, *Formula for Stress and Strain* by Raymond J. Roark. Similarly, a program called REEP is derived from "Rapid Electrical Estimating and Pricing" by Kenneth Kolstad. This type of program simply allows the user to locate his desired procedure in the book and then turn to the computer for the solution.

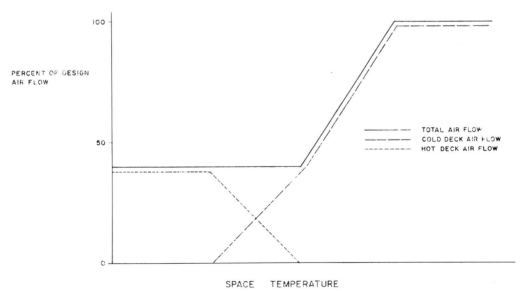

Fig. 38-2. Operation of Variable-Volume Double-Duct Mixing Box

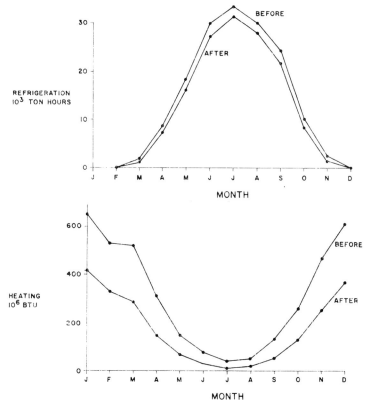

Fig. 38-3. Estimated Effect of Retrofit of Variable-Volume Double-Duct Mixing Box on Requirements for Refrigeration and Heating

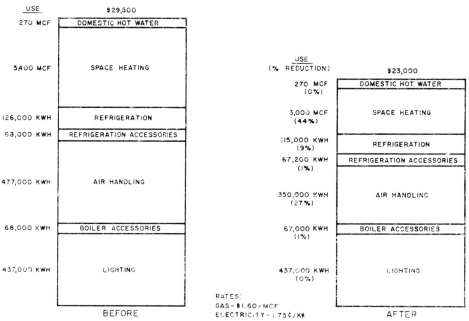

Fig. 38-4. Annual Energy Consumption Comparison of Variable-Volume Double-Duct Retrofit Community College Library

Engineers are good programmers

Once a design office has access to the high-capacity computer through the time-sharing network, inevitably, the day-to-day activities will generate uses other than the commercially available programs, particularly as the staff develops their programming skills. It might be mentioned at this point that by far the most valid engineering programs are generally written by skilled engineers who learn to program, rather than skilled programmers who try to understand the necessary engineering skills.

In this regard, the computer must be credited more than any other singular entity, with forcing the engineering community to address the systematic structure of the algorithms which dictate the behavior of engineering systems. The computer is a brutally specific device; it has been likened to a recalcitrant child, who does exactly what is demanded of him—no more, no less. It is therefore necessary to plan a computer approach to a problem much more specifically than a manual series of calculations. Before writing a program to aid in a specific design, the designer must know exactly what he wants to calculate and have his plan of attack thought out. Even when using an existing program, the designer is forced to be extremely specific in many details which he took for granted in prior efforts.

Time-sharing examples

Some examples of programs developed by an in-house engineering staff for use in a time-sharing mode include the following:

• *Utility rate schedules* are difficult to generalize, but a specific schedule can be handled by a calculator. It is just a short step further to program these in Fortran or Basic. Even a small number of calculations can be done effectively this way, and the computer output is its own documentation. Hand calculations are certainly a useful tool, but answers are rarely printed, and there always remains the opportunity for error in repetitive calculations. The advantage of having this programming capacity is, again, the virtually instantaneous turnaround time. The energy consumption output of a system analysis or so-called energy program can be interfaced with the rate program to give monthly and annual energy costs. Subsequently, by initiating an additional run (to determine the effects of a building design modification, alternative psychrometric system impact, or machinery module changes), the results on the annual energy costs will be immediately available.

• *Unusual system studies* for which available programs have not been developed are readily addressed. One slightly complex program was recently written for the purpose of evaluating the requirements for solar collectors. Many manufacturers give efficiency for collectors as a function of insolation and temperature excess at normal incidence or in a few cases, continuous performance curves at fixed latitude and plate angle. In the former instance, performance drops off at sun angles away from normal incidence, so that collection from sunrise to sunset is difficult to estimate. In rough terms, a collector cannot use any energy which has not come through the glazing. Using the equations in the *1972 ASHRAE Handbook of Fundamentals*, it is possible to calculate sun angles throughout any day of the year. Having these and the power series coefficients for transmission also given in the handbook, the transmitted radiation for any number of sheets of double-strength window glass can be obtained. Losses can, for a first approximation, be considered dependent on plate temperature alone and can be represented as a constant fraction of the energy collected. Figures 38-5 and 38-6 show the results of two such collector calculations, Fig. 38-5 giving monthly variation of solar gain, and Fig. 38-6 giving the hourly variation for two different months. Figure 38-7 superimposes hourly solar gains on hourly requirements for heating for the building considered.

• *Curve fit and regression* for a multitude of phenomena studied in the design of building systems can be handled readily when doing, for example, a psychrometric calculation (say, dry bulb and wet bulb temperatures input, dew point, specific humidity, relative humidity, and enthalpy output). It is unhandy to have the steam tables entered into the

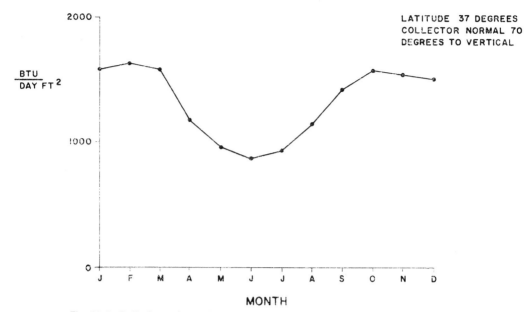

Fig. 38-5. Daily Solar Gains Through Two Sheets of Clear Double-Strength Glass

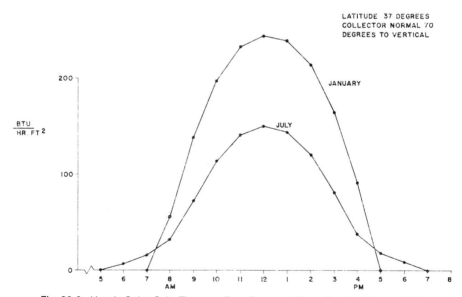

Fig. 38-6. Hourly Solar Gain Through Two Sheets of Clear Double-Strength Glass

computer so a simple mathematical equation can be much more easily used. Frequently, the user will have to develop the equations himself. Least-squares fitting techniques with canned or homegrown programs give a range of choices, along with calculations of "goodness-of-fit."

Figure 38-8 shows a few of the types of functional relationships which can be applied with least-squares techniques. The standard error of estimate is one measure of the quality of the fit.

Similarly, when doing a rough evaluation of the fuel consumption of a total energy plant, a

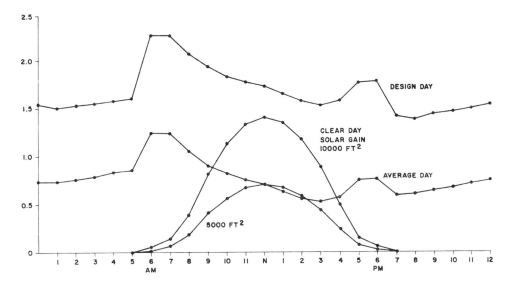

HOURLY HEATING AND SOLAR
COLLECTION PROFILES

JANUARY

Fig. 38-7. Prosposed Solar Assisted Heating System Dormitory Quadrangle

Two variables	$W = A + BX$
	$W = A + BX + CX^2 + \cdots$
	$W = A + \dfrac{B}{X}$
	$W = AX^B$
	$W = Ae^{BX}$
	$W = A + B \ln X$
Several variables	$W = A + BX + CY + DZ + \cdots$
Standard error of estimate	$= \sqrt{\dfrac{\sum\limits_{n=1}^{N} (\text{measured values} - \text{fitted values})^2}{N}}$

Fig. 38-8. Curve-Fitting Programs Typical Forms

multivariable regression allows fuel consumption to be expressed as a function of kilowatt hours generated and heating and cooling degree days and shows the deviations of the predicted from the actual consumption. Figure 38-9 shows the magnitude of the independent variables and the fuel consumption fitted. The maximum deviation between the measured and fitted values is 3 percent.

Another example seemingly does not relate directly to building environmental systems but illustrates how the computer can improve the operational success of energy systems design. In a gas-fueled total energy plant, the engine fuel consumption was excessive, but when adjustments were made in the carburetion linkage to optimize fuel efficiency at full load, the part load consumption was excessive. Figure 38-10 diagrams the problem. The lower part of the figure shows a schematic of

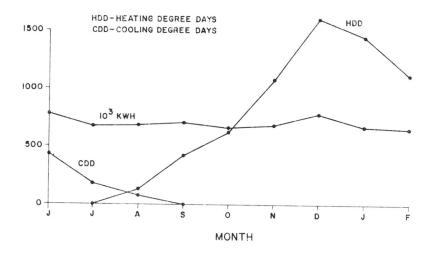

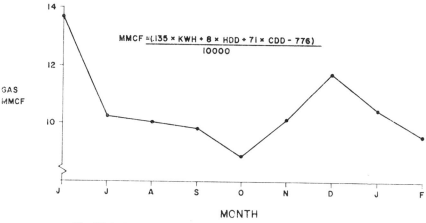

Fig. 38-9. Monthly Gas Consumption—Total Energy Plant

the linkage. The upper part indicates three relationships between air and gas metering valve positions. Curve I is the factory recommended setting. In practice, the engine will idle with the settings at point 1 but there is insufficient air flow at full throttle, point 2. If link B is shortened, as recommended in the engine manual, the air flow is increased at full throttle to point 3; however, the resulting air–gas valve curve is II, which has far too much air at idle (point 4). A program was written including the relationships of the linkage kinematics and the fuel–air ratios. The program output provided the exact adjustment data for the linkage elements, and the fuel consumption predicted was thus achieved. By properly lengthening link C and shortening B,

it was possible to obtain curve III, which has proper air–gas ratios at both idle and full load.

Optimum equipment selection

Another area of interest to the systems designer relating to computer technology is that of selecting the optimum equipment to integrate into the system. Currently, such equipment selection programs are available for many devices through the manufacturers. One frequently needed but particularly onerous calculation is the performance of a dehumidifying coil. Procedures have been somewhat standardized in ARI Standard 410-72, but they are typically iterative and can take a skilled engineer upwards of 15 minutes per point. Thus, coil selection programs are widely

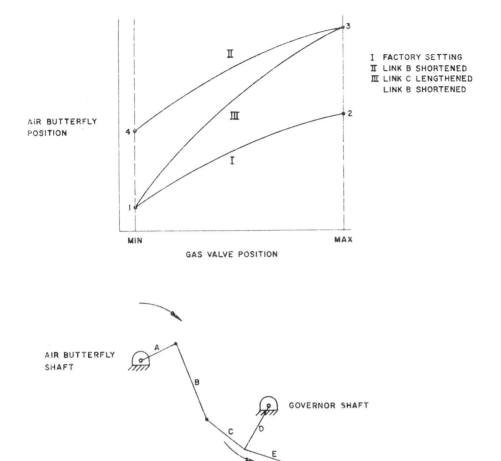

Fig. 38-10. Schematic of Linkage Fuel-Air Metering on Stationary Natural Gas Engine

available through coil manufacturers for selecting from their product line. Typical programs provide the designer with a range of coil selections to meet their psychrometric performance parameters. Each selection gives the coil dimensions, fin spacing, tube size, tube depth (rows), circuiting, water velocity, air side pressure drop, water temperatures, flow rates, and water pressure drop. With this information, the designer can intelligently select the coil that best satisfies all design parameters when integrated with his system.

Other equipment selection programs are available for devices such as chillers and heat exchangers. These programs were initially developed for optimizing the component match to minimize invested cost. However, with recent emphasis on total owning/operating cost and life-cycle costing, these programs have been expanded to include not only the investment cost parameter but the primary energy consumption, auxiliary energy burden consumption, and maintenance/service requirements.

Current state of the art

As one considers the current state of the art, it becomes evident that the application of computer technology is in its embryonic state in this field. Perhaps the most significant growth will proceed along with the anticipated surge in data communications technology. It is for this reason that the future of the computer for engineering calculations lies in the concept of a communication terminal in the engineer's office giving him access to information and programs either on shared-time networks or in the engineering departments of the component manufacturers, or both. In addition to a natural growth in the scope of the programs and concepts discussed, it is inevitable that in the future virtually all manufacturers' catalog selection data will be obtained from the computer terminal. This will provide the designer immediate access to the optimizing data, will allow him the advantage of iterating until he has optimized the selection of each component, will provide printed documentation of each selection, and it will provide the manufacturer with a continuing record of the activity his products are receiving in each office, allowing his sales engineering staff the opportunity to magnify their effectiveness by providing services more valuable than simple equipment selection.

SECTION VII

Financial considerations

The study of financial considerations related to engineering decision making is commonly referred to as engineering economics. This subject has been rather well structured, a number of excellent texts have been published on it, and courses in it are offered in most engineering curricula. Thus, it is not intended in this publication to provide principles or fundamentals of engineering economics.

In the building industry, unfortunately, there has been a serious lack of understanding and communication between the responsible design engineers and the owner agencies. Although the designer may be fully skilled and educated in engineering economics, the owner agency generally does not look to the engineer for participation in the financial decision-making process. As a result, the decisions are made independently of the engineer or his input, and the results of these decisions are provided to the engineer as design constraints. These constraints often take the form of forcing the first cost down to an unreasonably low level, dictating energy sources, etc., all of which may not be to the long-range benefit of the building investment.

The owner agency concerned first and foremost about a wise investment has no intention of making major financial decisions without the benefit of all the helpful input he can obtain. The unfortunate situation is that this segment of the building team has never been made aware of the tremendous impact that decisions made by the designer can have upon operating costs, and consequently upon the financial viability of the entire building venture.

Numerous major building developments in recent years have failed financially as a direct result of insufficient cash flow resulting from unanticipated energy costs or energy systems maintenance costs. Many others which are owned by public or private institutions would have failed had the buildings had to survive in the commercial money market. It was such a commercial project financial failure that stimulated the author to develop the bidding procedure discussed in two chapters in this section.

In the majority of these failures the designer has unfortunately and unjustly been criticized, when the true fault lies with the institutional structures of the building industry. Perhaps the exponentially rising costs of energy since 1973 will have a beneficial effect in this regard. Most building owners have begun to realize that energy costs are not simply a business cost based upon statistical norms, but rather, they are a truly manageable cost item, and that the first step in this management chain for a new building is to design to a given energy consumption. This is a concept heretofore that was virtually unthought of.

If this new era of communication comes to pass, the engineer must be prepared to participate. Many earlier efforts of some engineering practitioners to create an involvement in the financial aspects of the projects have failed rather completely, with the disappointed engineer being convinced that the investor or owner agency was not interested in his involvement. The fact of the matter was that the engineer was armed with his textbook knowledge of engineering economics, but totally lacked an understanding of the business methods of the building owner.

There are an untold number of ingredients in the economic formulas relating to buildings, and these are different for the many different classes of buildings. As an example, in commercial buildings, the variable ingredients include investment tax credits, corporate or individual income tax structures, cost of money, availability of equity, alternative uses for available funds, and many others. In many institutions, there is little impact from tax structures, but the invested monies and operating funds are totally unrelated sources—thus expenditures in one area to save in another cannot be justified!

This complex problem has been worsened by the promulgation of concepts and data on techniques called, popularly, "life-cycle costing" by numerous federal agencies. Many of these concepts fail totally to address the complex problems stated above and deal in theoretical or statistical monetary terms which only a governmental agency can accept as valid. The engineering community is not properly informed of the limitations of these concepts and, as a result, suffers more frustration.

The key, then, for success in the communication between the engineer and the owner agency is for the engineer to first inform the owner of his ability to optimize between investment costs and straight cash flow operating cost, and second, to obtain all possible information on investment criteria from the owner, then, finally to develop a technique to enable him to satisfy the owner's criteria.

When the chapter "Investment Optimization" was originally published, the author received numerous critical comments from economists and other proponents of more sophisticated investment/return techniques. Suffice it to say, that from the standpoint of economic theory or practice, the straight payback procedure used in the technique is not valid in and of itself. But the fact is that the owner agencies for which this purchasing technique was used were all able to ingest all of the complex variables relating to their specific investment, and to express the output *in terms of equivalent straight payback*! Had the author tried to become the financial analyst and have the satisfaction of handling all of the pure economic data (tax base, availability of funds, etc.), the benefits of the experiences related in these chapters would likely not have come to pass.

The message which is paramount in this section is that the engineering practitioner must learn and understand everything he can about economic considerations; he must instill confidence in his knowledge of financial matters with the building owner agency, and he must then accept whatever data that agency can provide and incorporate *that* data into the design parameter formulas. Expressed more briefly—he must be flexible in his approach to the financial parameters.

39

Energy-effective machinery can be self-financing (a case history)

Life-cycle costing is an effort to quantify the total owning costs of a system, subsystem, or component. Owning costs, in turn, can best be segregated into three basic components: 1) first cost or investment, 2) energy costs, 3) maintenance and service costs.

For many reasons the trend in product development in recent years has been to minimize first cost at the expense of the other two. However, with the recently accelerating increases in energy values and consciousness of the significance of maintenance and service costs, the trend is moving toward giving due consideration to these two components.

Life-cycle costing revealing

A (mid-1975) experience in life-cycle costing was most interesting and revealing. In an existing plant, two absorption water chillers were to be replaced with two new centrifugal chillers. The preliminary designs were completed, including development of flow diagrams and complete detailed load and energy studies. At this point, it was decided to request proposals on the chillers, and specifications were written for life-cycle proposals.

To account for energy costs, the following energy cost impact features of the machines were considered:

• *Power impact*—From the refrigeration part load analysis, it was determined how many months a year both chillers would operate, how many months one chiller would operate, and how many months no chillers would operate. A formula was then developed to provide a single number multiplier taking into account these demands established, the local utility demand rate, and present-day value of future dollars based on the owner's economic variables. The bidder would then simply multiply the full load compressor power (kW) requirement of the proposed unit by this compressor power multiplier to establish the power cost impact for the unit.

• *Compressor energy impact*—From the combination of the refrigeration energy analysis and the part load analysis, a formula was developed, linearizing the part load performance of the units, which provided a single number multiplier to be applied to the compressor power requirement at a single stated point of performance at less than full load. This formula took into account the calculated number of operating hours of each part load condition per year, the graduated commodity rate for the local utility company integrated with the other facility loads, and the present-day value of future monies as described above.

• *Water flow resistance impacts*—Similar single number multipliers were developed relating the pressure drops through the chillers and condensers, respectively, which considered the various parameters described, to be applied to the head loss for the chiller and product of head loss and flow rate for the condenser.

Include maintenance contact

With calculation methodology currently available, the energy and part load requirements can be quantified quite accurately. Anticipated maintenance and service requirements are, however, much more difficult (if not impossible) to quantify. Thus, it was decided that the only way to legitimately include this ingredient in the life-cycle proposal was to have each proposal include a full maintenance and service contract for an extended time by the manufacturer. The terms of the contract were specifically ennumerated in the specification.

Each manufacturer was then requested to submit a proposal on a stated series or model of machine to comply with specific performance requirements. The information given was full load system capacity, piping configuration, single machine capacity, chilled water flow rate, entering and leaving chilled water temperatures, entering condenser water temperature, and anticipated total machine hours of operation per year. Each manufacturer was invited to submit additional proposals on any alternative units in its product line that could meet the performance specifications.

Bid to include energy factors

The bid form included two sections; the first was product cost and energy impact. For any unit proposed, the bidder stated the cost of the machines and cost of start-up services; then entered the compressor full load kW, multiplied by the power impact factor; and entered the product as a power life-cycle value. Similarly, the compressor life-cycle energy value, chiller pressure drop life-cycle energy value, and condenser value were entered. These six figures were then added, giving the total life-cycle cost of the respective unit.

The extended maintenance and service contract proposal constituted the second section of the proposal form. This cost, by previously determined formulation, was then converted to present-day investment dollars.

Six proposals were received. The results are shown in Fig. 39-1. The ordinate represents the amounts and the abcissa the proposals, starting with proposal 1, the lowest life-cycle cost, and extending to the sixth proposal, the highest life-cycle cost. The top curve is the life-cycle costs and the lower one is the first cost, including start-up service. (The first cost was adjusted as needed to account for difference in installation cost between various ma-

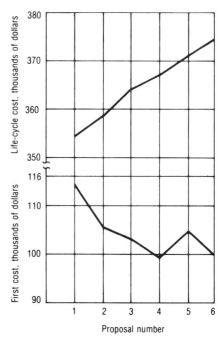

Fig. 39-1. Life-cycle and first cost versus bids received.

chines, such as insulation, field piping of ancillary devices, etc.) The impressive feature of the bids as revealed by the curves is the near-perfect divergence or inverse relation between first and life-cycle costs. It illustrates clearly that industry efforts at investment cost reductions have been at the expense of energy consumption and to a lesser extent, maintenance costs.

This methodology of purchasing major components, if not complete selection of systems, will go a long way toward quantifying decisions that were previously often made on the basis of judgment, but it is one more example of an ever-increasing maturity in the HVAC industry.

Author's Note: This chapter was originally published in December 1975. Subsequently, Chapter 40, which details the methodology, was published following additional experience with the technique. Approximately five years after the initial experience, the chiller manufacturers have responded by developing selection procedures to optimize the first cost versus energy-impact relationships of their units, and the divergence in the two curves shown in this chapter is becoming much less consistent.

40

Investment optimization: a methodology for life-cycle cost analysis

This chapter presents a methodology for life-cycle cost analysis that has been developed and subjected to marketplace experiences in both the private sector and state institutional sector.

Life-cycle costing (LCC) is a concept to which most practitioners in the building design profession readily relate. Much work has been recently done to promote the state of the art in this concept, the majority being funded and effected by various departments of the federal government. Dissemination of the data obtained, complex economic concepts employed, and experiences gained have been aggressively pursued both by the sponsoring agencies and the contractual manufacturers who have participated in some of the demonstration programs.

The fact remains, however, that the commercial and nongovernmental institutional sectors of the building industry—although anxious to take advantage of any available sound investment opportunities—have been slow to accept LCC techniques. As energy costs become an evermore significant component of the owning costs of thermal systems, and in keeping with tried and proven

This chapter was reprinted from *ASHRAE Journal*, Jan. 1977 and was originally entitled, "Investment Optimization: A Methodology for Life Cycle Cost Analysis." Appearance of this material in *Energy Engineering and Management for Building Systems* does not necessarily suggest or signify endorsement by the American Society of Heating, Refrigerating or Air-Conditioning Engineers, Inc.

business practices, the need for a sound methodology for applying those concepts increases.

Establish life expectancy

Most previous attempts at purchasing on the basis of LCC have not been successful because they were based upon a false premise. That premise was that one must identify the "life expectancy of the subject" to establish the LCC. To support this recognition, the fact is that life expectancy of such things as major machinery components, piping systems, building structures, and the like, is not a specific time span. It is, rather, one variable in a complex formula, dependent upon numerous other variables. The major independent variables, both interrelated with economics, are maintenance and obsolescence. As an example, a properly maintained refrigeration compressor will last indefinitely (barring destruction from "system" malfunctions), if the maintenance program established assures replacement of the wearing components prior to their total failure. If it is a nonmaintainable or nonserviceable compressor, then the compressor as a whole becomes a replaceable component of a larger subsystem (such as a chiller). Thus, in determining owning costs, if adequate monies are allocated for maintenance and service (M/S), the machine will never "wear out." Machinery and buildings are not organic objects such as animal and plant life—thus, man has total control over their "life expectancy."

Consider obsolescence

The second independent variable is the only one which truly limits the life of the building and the building system: obsolescence. The subject becomes obsolete when a more desirable subject is available to perform the same function.

Thus, to establish a true LCC study on a building (and its energy systems) the only realistic "life" that can be established must be based upon the estimate of the time preceding obsolescence.

The proposition, then, is that decisions made in the expending of funds to construct a building are not actualy made on the basis of the life expectancy of the structure erected. They are made on the basis of return on the investment in the case of commercial ventures, resolution or repayment of construction or corporate bonds in large businesses and institutions, and optimum cost/benefit ratios in state and local government institutions. Other bases sometimes exist, but most relate in some way to the free enterprise concept of return on investment (ROI). *The time preceding anticipated obsolescence, in this regard becomes the upper limit.*

To expand this proposition, then, perhaps the term life-cycle costing—when applied to monetary investments—should be changed to something such as investment optimization, or otherwise redefined to remove the stigma of direct life expectancy of the subject. Every astute businessman or investor is concerned with the solidarity of his investment for its life span, but his first and foremost concern is the economic wisdom of the investment! Thus, to interest such investors, the methodology employed will succeed only if it results in a more favorable investment.

Back to basics

In the development of the methodology, some basic economic concepts must be addressed.

First, no component of a building or building project "earns" money. To the contrary, virtually every component costs money to purchase and install, and thereafter continues to cost money to operate and maintain. For example, when a building investor purchases an air handling unit, after paying for it, he continues to pay to operate, maintain, and service it for as long as *he* owns it. What earns money is the space which is usable for a purpose. In an office building, the income is generated by the usable area, in a hospital it is the patient beds, laboratory products, etc. In both investment and operating costs, the component is simply one of the costs in providing the usable or revenue-generating area, beds, or whatever. The same is true in industrial production: The product earns the revenue; the economic complexities of the production machinery simply represent costs. Every investor recognizes this since the difference between sale price and costs either represents profit or improved service.

Thus, the familiar pro forma, or revenue-to-costs projection format is not the form to employ in developing an investment optimization analysis. The analysis assumes that either wisdom of the investment has already been confirmed on the basis of approximated or anticipated goals, or that the analysis is being employed to validate the pro forma.

No absolute best decision

Second, there is no "absolute" best decision in the purchasing of a component of a revenue-generating entity if the component has both investment cost burdens and operating cost burdens. Rather, there is only a comparative "best." For example, the decision to build a commercial office building, school, hospital, etc., is made on the basis of statistical economics, one parameter of which is "need." Once this parameter is tempered with other parameters such as the cost of investment monies, the degree of the need which reflects in possible rental revenue, tax parameters, and rough estimates of operating costs (energy, maintenance, insurance, taxes, etc.), the preliminary decision to proceed with the project is made. It is after this that the specifics of addressing the individual design decisions regarding systems and components are manifest. At this point, all decisions are comparative—System A versus System B, Unit C versus Unit D, etc. By resolution of these decisions, the value of the overall investment is optimized.

Third, the most beneficial or optimum investment is not the same for all investors or for all projects. Failure to recognize this is the paramount reason that the statement has been made ". . . *life cycle costing, as applied to the building industries, is relatively unaccepted.*"[1] Most techniques of LCC previously presented endeavored to key the formula into the life expectancy of the building or the component being analyzed. When, in fact, to the specific investor for a specific venture, this has little to do with his concerns.

Investment optimization method

A life-cycle costing methodology based upon the foregoing observations, entitled investment optimization (IO) is suggested. The first step at developing an IO analysis is to establish the requirements for the ROI on the basis of straight payback. The investor must, after taking all of the relevant parameters into consideration, reduce his expression of the invested dollar to a straight payback constant. This constant (n) is dimensionally "years." The formula is

$$I_d/R_d = n$$

where

I_d = differential investment (dollars),
R_d = differential return (dollars/year),
n = payback period (years).

The ratio n, is a function of many complex economic parameters. These include interest rates or value of investment monies, present value of future monies, tax parameters relating to capital investment versus operating costs, availability of investment monies, etc. Experience in the commercial and institutional building markets has shown that virtually *every* investor on *every* project can relate all these complex economic variables to the straight payback formula. (This is not to suggest that more sophisticated approaches such as "Rate of Return Method," "Discounted Cash Flow," etc. are not valid, just that once the decision is made, it be reexpressed in terms of straight payback.

Once the payback period "n" has been established, all investment associated terms of the owning cost have been removed from the formula. This feature facilitates the accomplishment of the investment optimization with a degree of simplicity which enables it to be used for all systems decisions and the component purchasing decision—to the extent that competitive bids can be submitted on the basis of investment optimization by applying a relatively simple bidding procedure. The steps in the methodology (including the first step above) are:

- Establish the straight payback period (n) that satisfies the economic parameters of the investor and specific project.
- Determine *all* operating costs associated with the system or components being considered for each year of the n year period starting with the first year. (*Note*: Interest, debt service, and depreciation are not a component of operating costs, they are taken into account in the step above). Operating costs include energy, maintenance/service, insurance, taxes (if relevant), and any other direct cost outlay related to the component or subsystem being analyzed.
- Add the initial purchase cost and the sum of the annual operating costs for the period of n years, for each of the systems or components being considered.
- The lowest total of the compared systems or components will provide the optimum investment.

As an example, consider the comparison between two competitive products, A and B, with first costs of $100 and $80, respectively, and annual operating cost burdens of $20 and $25, respectively.

Example 1. Investor has a requirement of a straight payback ratio, "n," of 3 years on invested monies:

	Product A	Product B
Investment Cost	100	80
Operating Cost		
1st Year	20	25
2nd Year	20	25
3rd Year	20	25
Total	160	155

The most favorable purchase would be product B. To substantiate the wisdom of the decision, if the incremental difference in investment were compared to the incremental return:

$$I = (100 - 80) = \$20,$$
$$R = (25 - 20) = \$5/\text{year, and}$$
$$n = 20/5 = 4 \text{ years.}$$

Thus, the 4-year straight payback is greater than the 3-year requirement, illustrating that the method used to select product B was valid.

Example 2. If the same products were compared for an investor who required a straight payback of 5 years:

	Product A	Product B
Investment Cost	100	80
Operating Cost		
1st Year	20	25
2nd Year	20	25
3rd Year	20	25
4th Year	20	25
5th Year	20	25
	200	205

In this case, the most favorable investment would be product A. As shown above, the straight payback rate of return would be 4 years, which is less than the 5 years which was established.

The costs

Investment Cost: Includes all actual costs (purchase and installation of component or system). Not included are interest costs and investment tax credits. If the component being considered is a water chiller, cost consideration would include chiller, freight, sales tax, contract cost to move into place and connect piping and electrical services, and insulation of components. Thus, *any* cash outlay associated with purchasing and installing the unit

that could possibly be different than any cash outlay associated with purchasing and installing a competitive unit. It is mandatory that the bid form analysis not use comparative costs but rather actual total costs.

Operating Costs: Include all conceivable cash flow burdens related to the specific component or system being analyzed, i.e., energy, power (demand), maintenance/service, insurance, and taxes. Any other item that relates should be included. Each item should be calculated separately. For more accurate analysis, anticipated escalation or inflationary increases should be taken into account. Also, economic analyses reveal that different escalation rates should be applied to various cost items as the case warrants, i.e., energy cost inflation may be anticipated to follow a curve different than M/S cost inflation. For each item of cost, then, an escalation rate unique to that item can be applied. The escalation rate formula is simply:

$$C = c_1 \left(\frac{(1 + x)^n - 1}{x} \right)$$

where

C = total cost of the operating component over n years;

c_1 = annual operating cost component for first year;

n = straight payback ratio;

x = annual (compounded) escalation rate (10 percent = 0.10).

Energy and Power: These costs are probably the easiest major operating cost component to quantify with a reasonable degree of accuracy. Systems analysis techniques that relate to energy and power burdens for systems and most system components have been formulated in computer programs that are commercially available to design professionals. These programs can assist the designer in determining maximum machinery loads per month (demand), energy input per month, operating hours per month, and part load profiles. These data can then be used to determine power and energy burdens for any specific machinery component or integrated system. For bidding purposes, the data can be integrated with the applicable energy rate or fuel costs and reduced to a series of constants to be used as multipliers for various energy burden characteristics of the machinery being compared. For example, for a centrifugal water chiller, constants would be generated for power costs by multiplying full load power input for the specific unit, by a constant to produce the "n" year costs. The part load analysis is integrated with a linear approximation of part load power curves to be multiplied by the power input at some relevant fraction of part load to produce the "n" year primary energy cost. The evaporator flow rate incorporated with either the total hours of operation (constant flow), or part load analysis (throttled flow control) is integrated with energy cost parameters to develop constant to be multiplied by the head loss through the respective evaporator to produce the "n" year chiller pumping energy burden. The operating hours analysis integrated with energy rate parameters is used to develop a constant to be multiplied by the product of the required flow rate and head loss through the condenser to produce the condenser pumping energy burden.

Maintenance/Service: Costs associated with a particular item of machinery or system are the most difficult to quantify accurately. Most published data are statistical-historical in nature have not been correlated with specific machinery. ASHRAE Technical Committee 1.7 (Durability, Reliability, Maintainability) is currently working on the development of a methodology to enable the industry to address this problem.[2] Until some progress is made in this effort, the only reasonable method of incorporating M/S in the IO formula is either for the design professional to perform a thorough investigation on typical classes of components when doing the analysis on comparative system; or, when applying IO to bidding, to require bidders to include a total maintenance/service contract proposal for some years, including as a minimum, the "nth" year.

Insurance: As a general rule, the energy and M/S cost burdens are the major differential financial burdens of operating machinery. However, in some cases, machinery insurance rates will reflect the statistical differences in probability of loss or liability for alternative selections of machinery. Thus, prior to conducting an IO analysis or preparing IO bid documents, the investor's insurance underwriter should be consulted to determine if any potential differences, in fact, do relate. For example, some premiums on electric refrigeration machinery are based on *motor* horsepower; thus if a smaller prime mover can be used for the specified tonnage, some savings could result. If such a situation exists, appropriate consideration should be given in the analysis.

Taxes (Real Estate): They are generally the least significant contributor to the differential operating cost. However, if any differences are identified, they should be considered. Most taxing agencies base the appraised valuation, not on actual cost, but on the revenue-generating capability of the venture or some statistical appraisal. If, however, actual costs are considered, an appropriate constant should be applied to the purchase price to reflect any such tax burdens.

The investor in Example 1 may be representative of the speculative investor for whom investment monies are costly, and the investment is short term. If the straight payback ratio is not skillfully applied, the investor could do himself a costly disservice. Consider that if the investor of Example 1 retains the property for, say, 20 years, with no escalation of operating costs, his differential operating costs at the end of the period will have been five times the savings in first cost. Thus, it is extremely important that the consultant advising an investor on investment optimization make him totally aware of the significance of the straight payback ratio established. Failure on the part of many nonsophisticated investors to recognize the importance of a realistic straight payback ratio has resulted in the economic failure of building projects through foreclosures, and the subsequent ownership of properties by investment banking houses who found it difficult to salvage the invested capital. On the other hand, all sophisticated investors fully understand the wisdom of increasing the "n" ratio to the greatest number possible when they anticipate so-called permanent ownership. Another germane consideration is that the anticipated time preceding obsolescence must be, as a minimum, equal to $2n$ years.

Case history examples

Case history examples of the IO methodology for LCC reveal a significant improvement in the decision-making process of purchasing major machinery under competitive bidding procedures.

• *Case 1*: Being considered were two water chillers for a commercial building development. Constants for the energy and power components were developed, following systems analysis, to determine the part load profiles and monthly hours of operation, and integrating this data with the local utility rates. For this bid, the straight payback ratio (n) was set by the owner as 5 years, and no escalation in energy was included because of a unique local condition relating to electric-generating capacity and fuel source. The "bid form" in a capsulated format is shown in Fig. 40-1.

A rather well-detailed specification on performance tests, and the maintenance/service cost was, needless to say, required. The results of the bids are presented graphically in Fig. 40-2. The lower curve is the sum of lines A and B, or the total first cost, and the upper curve is the total IO bid cost. Note that the incremental

first cost difference between the lowest IO bid cost units and the second lowest is $8750. The difference in annual operating costs is $1907. The straight payback ratio, in years, is then

$$8750/1907 = 4.59.$$

Thus the investor's ROI ratio has been satisfied. It might be mentioned, that if only the energy component of operating costs is considered, and present dollars are applied for a 20-year ownership "life cycle" or investment duration, the lowest IO bid compared to the second lowest will reduce the 20-year energy cost by $49,615. The same comparison made between the lowest IO bid unit (highest first cost) and the lowest first cost unit yields a 20-year energy cost saving of $87,472—not to mention the reduction in resource depletion.

• *Case 2*: Being considered is a single water chiller to be connected into a campus chilled water-loop system. The same procedure for establishing the analysis constants was employed as in Case 1, and again, the owner established a straight payback ratio, "n," of 5 years. However, there were four significant differences mandated by the specific or unique circumstances of the project and bidding procedures: (*1*) *The chiller was a component of a large integrated system. Thus, any capacity in excess of the specified minimum amount (385 tons) could be utilized in the system. A cost per ton correction was employed in the IO bidding procedure; (2) The IO procedure for bidding the chillers was structured as a set of alternates to the base bid.*

A. Sale Price—on delivery $ _____

B. Start-Up and Performance Test $ _____

C. *Compressor Energy*
     kW of two machines operating in series at 720 tons total
     load _____ × 266   = $ _____

D. Compressor Power
     kW of two machines operating in series at 1200 tons total
     load _____ × 79.6   = $ _____

E. *Chilled Water Energy*
     ft chilled water head loss through one machine
     _____ × 682.86   = $ _____

F. Condenser Water Energy
     ft head loss one machine _____ × GPM
     one machine _____ × 0.150   = $ _____

G. Five-year charge for total maintenance and service contract   = $ _____
     IO Bid Cost (Total A through G) $ _____

Fig. 40-1.

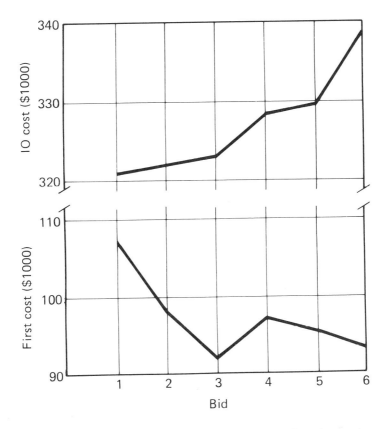

Fig. 40-2. Comparison of IO cost versus first cost for Case 1 example

Thus, the first costs were actually installed costs reflecting any machinery-related differences in contractor installation costs; (3) The owner was a state institution and could not legally receive bids for a five-year maintenance operations contract. Thus, the IO bids did not reflect differences in maintenance costs between alternative selections; (4) The energy

and power constants were developed on the basis of 10 percent per year compounded cost escalation.

The "bid form" for this project, in a capsulated form appears in Fig. 40-3.

The results of the bids are presented graphic-

Alternate No. X, IO Evaluation of Alternate X		= $ _____
A. *Weighted Sale Price Value*		
Quoted Price _____ × 385/Capacity _____		= $ _____
B. *Compressor Energy*		
Machine kW at 290 tons load _____ kW × 451.4		= $ _____
C. *Compressor Power*		
Machine kW at 385 tons load _____ kW × 102.5		= $ _____
D. *Chilled Water Energy*		
ft of heat pressure drop thru chiller at		
1400 GPM _____ ft × 246.5		= $ _____
E. *Condenser Water Energy*		
ft head _____ × GPM _____ × .176		= $ _____
IO Bid Cost (Total A thru E)		$ _____

Fig. 40-3.

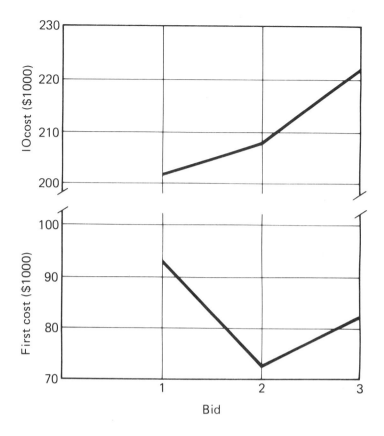

Fig. 40.4. Comparison of IO cost versus first cost for Case 2 example

ally in Fig. 40-4. Again, the lower curve is the alternate bid price or first cost, and the upper curve is the stated IO bid costs. Comparing the lowest IO bid cost to the second alternative, it is found that the quotient of the difference in investment cost between the lowest IO cost unit and the lowest first cost unit (second lowest IO cost) is:

$$\frac{\text{differential investment}}{\text{differential annual operating cost}} = \text{``}n\text{''}$$

$$20{,}820/7277 = 2.86 \text{ years.}$$

The additional investment, returned in 2.86 years (straight payback) will reduce the owner's energy costs compared to the lowest first cost alternative over a 20-year ownership life by $114,428.

In both cases cited, the owners elected to purchase the more costly machinery to obtain the desirable return on investment and the added benefit of considerable reduction in the so-called "life-cycle cost."

Conclusion

The IO methodology for life-cycle purchasing is a technique that enables commercial and institutional investors in buildings to obtain a realistic balance between the economic advantages of return on investment and the benefits of LCC. Although not necessarily achieving the lowest LCC as defined or addressed by earlier demonstrations, it does enable the earlier concepts to be applied to the real world of monetary economics of investment and return. Furthermore, it has the flexibility in structure to be applicable to virtually any economic and technical situation.

The use and validity of the IO technique has

been proved in numerous systems analysis decisions in the design process. Its simplicity has made it readily adaptable to competitive bidding procedures. The only requirements are the ability of the owner or investor to express his ROI requirements in straight payback terms, and the ability of the analyst or designer to quantify the primary and relevant cost variables of operation.

References

1. David Rosoff, "The Background, Progress and State-of-the-Art in Applying Life-Cycle Concepts in Building and Systems." Conference on Improving Efficiency and Performance of HVAC Equipment and Systems for Commercial and Industrial Buildings, Ray W. Herrick Laboratories, Volume 1, Purdue University, April 1976.

2. ASHRAE Handbook, 1973 "Systems," Chapter 44 "Owning & Operating Costs."

41

Single equation for cogeneration financial feasibility determination

The second law of thermodynamics tells us that a machine whose working fluid undergoes a cycle cannot absorb heat from a high-temperature sink and produce shaft work without rejecting heat to a lower temperature receiver. The obvious way to improve the effectiveness of use of the input (high-level) energy, then, is to put the heat rejected by the cycle to some beneficial use. From time to time, as economics, materials, machinery, and the energy needs of society change, these combined cycles are found to be of some advantage in the overall scheme of energy utilization.

In the early days of the twentieth century, combined or integrated plants were quite commonly used in closely knit urban areas to provide utility electricity and steam; such plants were also used almost universally to provide electricity and heat for large campus-type institutions. The Carnot principle, advances in materials technology, and monetary economics led the utilities to larger plants that condensed steam at temperatures too low for beneficial use. The problem of providing heat was thus separated from that of providing electricity at a time commencing generally in the 1930s.

With the era of relatively inexpensive natural gas and fuel oil in the late 1950s and throughout the 1960s, such plants again became an attractive vehicle for providing electricity, heat energy, and cooling for buildings. This "new" systems technology was called *total energy*, and most such plants utilized internal combustion engines rather than steam Rankine cycles. Increased fuel costs, increased

costs of money, and other circumstances (see Chapter 45) brought an end to the popularity or common use of such plants in the early 1970s.

Cogeneration is new term

The latest interest in the concept is appearing under a new label—*cogeneration*. Regardless of the name, the technique is valid when the correct conditions and circumstances warrant its use; and regardless of the validity of the concept, the technique is invalid when these circumstances do not exist.

Much time and expense can be avoided in the conduct of studies, analyses, and research if a simple sensitivity equation is applied to the problem as a first step. Such an equation, which applies to a system that consumes fuel and produces electricity and usable heat (Fig. 41-1), is presented and defined in Fig. 41-2. A very important limitation on the use of this equation is that it applies *only* to systems with the product shown in Fig. 41-1. If one of the

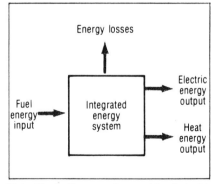

Fig. 41-1. Block diagram of a combined electricity/heat cycle.

A single equation that can be used for feasibility and sensitivity analyses of combined electricity/heat cycles is:

$$K_g = F(10^{-4})[R - (1/\eta_B)(H_rU)] + M + I + X$$

where

K_g = cost to generate electricity, ¢ per kW-hr
R = prime mover fuel rate, Btu per kW-hr
F = cost of fuel, $ per million Btu
H_r = salvage heat available, Btu per kW-hr
η_B = boiler efficiency in producing heat from fuel ($\eta_B < 1$)
U = utilization ratio for recovered heat ($U \leqslant 1$)
M = cost of maintenance, ¢ per kW-hr generated
I = amortized investment cost, ¢ per kW-hr
X = any other fixed costs, ¢ per kW-hr

The value of K_g determined from the equation is compared to K_p:

K_p = cost to purchase electricity, ¢ per kW-hr.

Substitution of parameter values developed in the text for the example problem is illustrated below:

$$K_g = F(10^{-4})[13,000 - (1/0.65)(4680)(0.75)] + 0.72 + 2 + 0.48$$
$$= 0.76F + 3.2.$$

The final expression is plotted in Fig. 41-3.

Fig. 41-2. The Cogeneration equation for electricity-heat cycle.

products is cooling (such as chilled water), this equation does not apply.

By direct application, one can readily determine the effect of any of the variables in the equation on the economic feasibility of the plant by changing that variable, holding all others fixed, and comparing Kg to Kp. When Kg exceeds Kp, it is not feasible to apply the combined cycle; when Kp exceeds Kg, it may be. The least controllable or predictable terms at this time are the costs of electricity and fuel. So it may be desirable to perform an analysis to determine at what relationship between electric and natural gas rates a plant might prove advisable.

Consider practical example

Consider an example. Kg is to be expressed as a function of F for a plant of 500-kW capacity, using the utility company as a standby for the full demand. The plant is to be a natural gas engine plant and will produce 2.5 million kW-hr of electricity per year. The variables for the equation are developed as follows:

• The plant will consist of two 250-kW engines, each with a fuel rate of 13,000 Btu per kW-hr at the average load.

• Average boiler efficiency for producing steam is assumed to be 65 percent.

• For the engine used, 36 percent of the input energy can be recovered as 10 psi steam. Therefore, $H_r = 0.36 \times 13,000 = 4680$ Btu per kW-hr.

• A comparison of the thermal and electrical load profiles reveals that 75 percent of the salvage heat can be used; thus, $U = 0.75$.

• Discussions with the engine manufacturer's service agency reveal that a complete maintenance and service contract can be obtained for $1.20 per running hour per engine. With two engines, each anticipated to run 7500 hr per year, the maintenance/service costs are $(7500 \times 2 \times 1.20)/2,500,000$, or $M = 0.72$¢ per kW-hr.

• The plant is estimated to cost $500 per kW, or a total of $250,000. The investor has determined that a 20 percent return on investment per year is acceptable; thus, the annual cash requirement for amortization is 20 percent of $250,000, or $50,000. I is then the quotient of the annual amortization cost and the annual kW-hr generated, or 50,000/2,500,000 = $0.02 per kW-hr or 2¢ per kW-hr.

• The utility company will provide standby

electric service at a monthly charge of $2 per kW of contract demand, with a resulting monthly charge of $1000. The value of X is then the total of the 12 monthly standby charges divided by the annual kW-hr, or 0.48¢ per kW-hr.

Substitution of the variables determined above into the basic equation is illustrated in Fig. 41-2, yielding:

$$K_g = 0.76F + 3.2.$$

This equation is graphically illustrated as a plot of K_g versus F in Fig. 41-3. In this way it can be seen that if the available fuel cost at the proposed site is, say, $3 per million Btu, cogeneration could be feasible if the average purchased electric *cost* is above 5.5¢ per kW-hr. If it is below this value, no further consideration need be given to the concept.

In the normal sphere of potential applications, such plants which produce electricity and beneficial heat energy only are impractical because of the extremely low utilization ratio for the byproduct heat. In a "normal" building systems application, this utilization

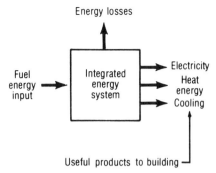

Fig. 41-4. Block diagram of an integrated energy plant providing electricity, heating, and cooling products.

will be approximately 18 percent in temperate climates. In an effort to improve the utilization, designers and analysts have sought additional uses for the power cycle reject heat. (It must be remembered that if the steam use is created simply to improve the *plant* economics, the financial burden will still fall upon the *building* economics; see Chapters 26 and 43.) One evident use is to motivate thermal cycle refrigeration to be used for building cooling; the ideal feature here is that the cooling and heating complement one another in the time cycle.

When this concept is incorporated into the cogeneration plant, it takes on the product configuration illustrated in Fig. 41-4. In performing a legitimate analysis, one *must* consider the cooling cycle an integral component of the plant, and the cooling product a plant output. This is true even if the cooling units are physically decentralized and not located within the plant, as on many college campuses.

Four terms added to equation

The cogeneration equation developed above must be provided with an additional term to credit the cost of generating electricity with the value of the cooling product. The equation shown in Fig. 41-5 gives the additional deductive cooling term. The inclusion introduces four additional terms and revises one term in the original equation.

To illustrate the use of the equation, consider the case of a college campus, first using the salvage heat for heating only and then adding the cooling component. The cost to

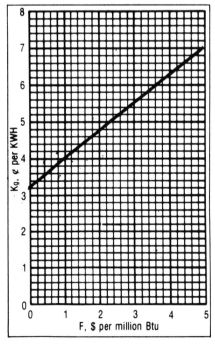

Fig. 41-3. Cost to generate electricity versus cost of fuel for example problem.

- For electricity and heating only:

$$K_g = F(10^{-4}) [R - (1/\eta_B) (H_rU_H)] + M + I + X.$$

- For electricity, heating, and cooling:

$$K_g = F(10^{-4}) [R - (1/\eta_B) (H_rU_H)] - (1/A_R) (E_RU_CH_rK_P) + M + I + X$$

where

- K_g = cost to generate electricity, ¢ per kW-hr
- R = prime mover fuel rate, Btu per kW-hr
- F = cost of fuel, $ per million Btu
- H_r = salvage heat available, Btu per kW-hr
- η_B = boiler efficiency in producing heat from fuel ($\eta_B < 1$)
- U_H = utilization ratio for heat recovered for heating ($U_H \leqslant 1$)
- U_C = utilization ratio for heat recovered for cooling ($U_C \leqslant 1$)

  Note: $U_H + U_C \leqslant 1$
- A_R = fluid heat rate for absorption cooling, Btu per ton-hr
- E_R = energy requirement for compression refrigeration, kW per ton
- K_p = cost of purchased electricity, ¢ per kW-hr
- M = cost of maintenance, ¢ per kW-hr generated
- I = amortized investment cost, ¢ per kW-hr generated
- X = any other fixed costs, ¢ per kW-hr generated

Substitution of parameter values developed in the text for the example problems yields:

- Heating-only arrangement; see Curve 1 in Fig. 41-6

$$K_g = F (10^{-4}) [46,667 - (1/0.75)(31,587)(0.30)] + 0.75$$
$$= 3.40F + 0.75$$

- Heating and cooling arrangement; see Curve 2 in Fig. 41-6

$$K_g = F(10^{-4}) [46,667 - (1/0.75)(31,587)(0.30)] - (1/18,000) (1 \times 0.25 \times 31,587 \times K_p) + 0.75$$
$$= 3.40F - 0.44K_p + 0.75$$

If K_p is set equal to K_g,

$$K_g = 2.36F + 0.52$$

Fig. 41-5. Congeneration equations for Plant Providing Electricity, Heating, and Cooling.

generate electricity is to be calculated in terms of the fuel cost. Assume that the college already has a cogeneration system that burns natural gas and coal and generates electricity with a Rankine cycle using steam turbines. The machinery cost has been amortized to a zero cost/value, but the maintenance cost covers day-to-day component replacement. This cost is determined to be 0.75¢ per kW-hr; thus, the only fixed cost component is M, which is equal to 0.75¢.

- The steam rate of the turbines is 35,000 Btu per kW-hr, and the average boiler efficiency is 75 percent. Thus, η_B is 0.75, and R is 35,000 ÷ 0.75 = 46,667 Btu per kW-hr.
- The recovered heat available is the turbine heat rate less the heat value of the kW-hr

generated; 35,000 − 3413 = 31,587 Btu of salvage heat per kW-hr generated.

- The utilization ratio for salvage heat used for heating, domestic hot water, and other building needs is 30 percent, or $U_H = 0.30$.
- The steam used for absorption refrigeration is anticipated to be 25 percent of that generated; thus, U_C is 0.25, and the steam rate for the absorption units, A_R, is to be 18,000 Btu per ton-hr.
- For comparative purposes, the refrigeration could be produced at 1 kW per ton if electric compression machinery were used.

The solution to the heating-only plant equation is shown in Fig. 41-5 to be $K_g = 3.40F + 0.75$. The analysis for the plant that includes the cooling can be performed in two different ways. The value of available purchased power can be inserted for the term K_p and the calculation performed to see if K_g is less than K_p.

As an example, for this analysis if it is determined that the fuel cost is \$2.50 per million Btu and that purchased electricity to drive electric refrigeration would have an average cost of 4.5¢ per kW-hr, the equation would reveal that the cost to generate, K_g, will be 7.27¢ per kW-hr; this, being higher than the 4.5¢ per kW-hr, indicates that at these fuel and electric costs the plant is economically unsound.

Determine economic feasibility

But to determine the economic feasibility under varying conditions of costs of fuel and purchased electricity, we can set K_p and K_g equal and determine the crossover point for economic viability. In the example, if K_g is set equal to K_p, the solution in terms of K_g and F is:

$$K_g = 2.36F + 0.52.$$

And it becomes immediately evident that if the fuel cost is \$2.50 per million Btu, the plant would not be economically viable unless the cost of purchased electricity were at least 6.42¢ per kW-hr.

The single-equation approach is intended as a preliminary analysis when consideration is being given to the cogeneration option for any type of plant or cycle(s). It must be emphasized, however, that such a simplified approach is applicable only as a first-cut analysis. It can be used to reject those opportunities that are not feasible, reveal the relative importance of the various input functions, and identify those opportunities that require further investigation.

Another interesting application involves the sensitivity aspect. As an example, studying the results as the utilization ratio varies reveals some of the fundamental economic differences between an internal combustion plant

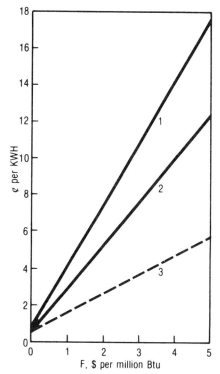

Figure 41-6. Cost to generate electricity versus cost of fuel for example problem; Curve 1 applies to heating-only arrangement, and Curve 2 applies to heating and cooling arrangement. Curve 3 defines hypothetical case with 100 percent utilization of heat by-product.

with a heat-to-electric ratio of 1.5 to 1 and a Rankine plant with a ratio of 9.25 to 1. It can be seen that the economics of the Rankine plant are vastly more sensitive to the utilization ratio. At full utilization conditions, there is little difference in the efficiency of source energy utilization for any type of well-designed cogeneration plant.

For illustrative purposes, Curve 3 in Fig. 41-6 illustrates what the values of K_g would be if U_H were increased to 75 percent, making the total utilization ($U_C + U_H$) 100 percent. It should always be kept in mind, however, that this is not economically valid unless there are true and beneficial end use needs for all of the heat. It was this misunderstanding in concept that was the topic of discussion in Chapters 26 and 43.

SECTION VIII

Energy source systems

The energy source or sources for a building can be thought of as meaning one of two things: either an initial source in its form as found in nature, or the form in which a central conversion plant in the building, or remote from the building, delivers it to the use system or some intermediate transport system. In the final analysis, studies of the building source system must include consideration of both the initial natural resource which is employed and the method in which that resource is converted to the ultimate building use form. Lack of recognition of this fundamental principle has led to misdirected source selection in a significant percentage of the buildings constructed during the period commencing about 1945, and if not recognized in the future, will continue to plague efforts to address the energy source selection intelligently.

In the era of aggressive energy marketing during the 1950s and 1960s, little attention was given to the initial resource from which the energy was obtained or the method or efficiency of converting it. The primary consideration was monetary economics which themselves were often misapplied. This situation was primarily existent in North America where the energy supply industries had been geared up for the needed production of World War II and were seeking new markets, supported by a combination of available resources, technology, and capital. It has only been since the mid-1970s when the age-old economics law of supply and demand started exerting pressure in the marketplace that both the owners of buildings and the designers of the energy systems have become concerned about energy on other than a present-day economics basis.

Unfortunately, when a particular parameter becomes a popular issue, it tends to be applied out of proportion to its true relationship to the other parameters. The first chapter in this section is simply a checklist aimed at keeping the energy source/system selection in perspective. The chapter on the "Myth of Free Steam," illustrates by simple but hopefully valid arithmetic the error of not considering the time-integrated nature of energy consumption and the error of not reexamining yesterday's technology today. In an earlier era, when alternative energy sources were not available to serve relatively large loads, the builders of these large facilities found it necessary to construct their own conversion plants. It was thus that many institutions and large buildings contained electrical generating plants. Since state of the art at that time was primarily turbine–Rankine plants, the concept of utilizing turbine exhaust for building heat was a viable method of increasing the effectiveness of energy utilization during some months of the year.

163

We now live in a different era and must reevaluate these prior decisions in view of new technologies and economic structures. In the combined power-heat cycle of the steam Rankine system, the steam is first expanded through the turbine, then condensed in either a heating system or for some beneficial process heat. Conceptually, this source cycle can be shown to have a design value (power) "efficiency" of 70 to 80 percent. It can also be shown that *each* of the products produced (electricity and heat) is produced at the combined cycle efficiency. This appears to be the only way to provide shaft power from heat at higher than the Carnot efficiency. Thus the reason for all the attention to the so-called cogeneration concept in the late 1970s.

The shortcoming with most applications of the cogeneration concept is that the whole problem has not been addressed—some of the ingredients of which are the identification of valid thermal loads, the load balance between shaft and thermal needs versus the machinery balance between shaft and thermal capability, and the time-integrated variations in both the absolute quantities of the loads and the balances. Only when these other parameters are properly analyzed, can the true wisdom of specific cogeneration selections be determined.

There was a considerable amount of activity in the decade of the 1960s in what is now called cogeneration. The popular thrust at that time utilized internal combustion (Otto or Diesel) cycles. Chapter 44, "An Oil-Fired Integrated Plant Design" is a step-by-step discussion on the design approach for such a facility. Although we are currently reconsidering the wisdom of utilizing oil for building systems, most of the materials in that chapter are also applicable to the design of natural gas fired plants. The final chapter in this section reflects upon the status of this internal combustion cogeneration activity (called total energy) at the time of its original publication in May 1978.

One statement in the first chapter might be discussed in more detail, as it is probably the single most important consideration in the selection of a building's energy source. That statement is as follows:

Reliability of the source, however, is an often overlooked parameter. If there is any question of future availability or future stabilized cost, the system should provide for conversion to an alternate source.

In designing the source system, this simple statement should never be overlooked—doing so could destine the building to an early and unanticipated obsolescence. This is particularly true if the resource source is not readily identified as is the case when the building receives a preconverted form such as electricity or district steam.

42

Selecting an energy source and conversion system

Many decisions are required as the design of a building thermal environmental system develops. This chapter addresses some suggested parameters to apply in the selection of energy sources and primary conversion systems.

Four energy forms available

The state of the art in available hardware has resulted in two forms of energy being needed to motivate and supply the HVAC needs of a space; these are thermal energy and electrical energy. The forms available for consideration for building systems as supplied are: electricity, fossil fuels; district heating and cooling, and solar energy.

Although there is some interrelationship between the use form and the supplied form, it is quite minimal. For example, available technology dictates that electricity is universally used to provide the energy for auxiliary system drives, such as fans and pumps for any number of uses, leaving the interrelationships between use and source to the primary motivating energy of the primary high-level (heat) and low-level (cooling) source systems. Thus, except for these two selections, the remainder of the subsystems are best designed irrespective of the source.

Energy source parameters

The selection of the building primary conversion system, however, cannot be separated from the source. The suggested parameters are:

- *Availability of source matter or energy—* It would be ridiculous to consider natural gas in an area where it is not available (a rather obvious parameter).

- *Unit cost of matter or energy*—Probably the most obvious parameter, and one which many times has been considered without regard for any other parameters.

- *Integrated efficiency of conversion to the required use form*—The term integrated is of paramount importance since the building energy system is a time-integrated entity; and over a given time span (such as a calendar year), it operates at less than design load the vast majority of the time. This realization coupled with the inherent reduction in efficiency at less than full load mandates consideration of the time-integrated efficiency.

- *Investment cost of storage, handling, and conversion apparatus*—Like the energy unit cost, this parameter appears obvious. However, in this context, it is emphasized that investment cost considerations of other subsystems, such as *distribution and terminal conversion or control, are to be considered separately.*

- *Environmental considerations of the space*—This is an indirect parameter in that it relates to a consideration of the basic need for energy. For example, in a warm climate, if thermal source energy is being considered for the purpose of dehumidification only, perhaps alternative methods of satisfying that need could be developed. Another example might be the reconsideration of utilizing high-level lighting heat when a fuel source might prove more energy resource effective.

- *Environmental considerations of the community*—Although it may seem a bit idealistic to suggest considerations beyond statutory requirements, such idealism has relevance both morally and potentially economically. From the moral side, all engineers are sworn to consider the impact of their designs on the public welfare. Economically, future legislation could have significant financial impact on a building's primary conversion system (whether on- or off-site) if the system is considered detrimental to the environment.

- *Energy consumption and demands (power)*—This parameter, coupled with the unit cost schedule, is required to determine the ultimate annual comparative energy cost.

- *Cost and availability of maintenance and service for the conversion apparatus*—Although this may appear to be an obvious ingredient to those who support the concept of life-cycle cost analysis, it is the least easily quantifiable ingredient thereof. The hourly cost may be easily identifiable, but the hours required are most difficult to assemble or predict. The second consideration, availability, is the most overlooked parameter in energy conversion system selection. The availability of skilled maintenance and service personnel is highly regional in nature, particularly for sophisticated machinery and controls.

- *Cost and availability of replacement components*—Keeping in mind that most buildings are long-term investments, the conversion apparatus and its controls must be selected with the consideration that replacement components will be readily available for years to come.

- *Reliability of the source and conversion apparatus*—Reliability of the conversion apparatus, whether it be a boiler or a transformer/switchgear combination, is interrelated with cost; i.e., redundancy could achieve the needed reliability. Reliability of the source, however, is an often overlooked parameter. If there is any question of future availability or future stabilized cost, the system should provide for conversion to an alternate source.

Not included are sources dictated by code or legislation. This is not an oversight, but rather an expression of faith in a free economy. Hopefully, well-founded application of design parameters to energy source selection as well as other building conversion systems will result in the most effective use of energy resources; such that if and when federal or regional policies affecting energy use are adopted, the engineering profession will have arrived there first.

43
The myth of free steam

The concept of getting something free, without paying a price for it, has fascinated man for as long as he has existed. Classical and commonplace philosophers, however, throughout history, have recognized that, as it is expressed in today's commonplace terms, "there is no such thing as a free lunch." Works expressing the fallacy of something for nothing span from Aesop in 550 BC through the *Arabian Nights* tale of Aladdin and Shakespeare's works to Browning's *Pied Piper* in the nineteenth century. There are countless examples.

Unfortunately, leaders in business, government, and even some decision-making members of the engineering community tend to be misled by the something for nothing concept as it is related to energy conversion systems. Understandably, we must continually seek more effective methods of converting our limited energy resources to the needed forms in the needed quantities required by society. However, in striving for our goals, we must be most careful to avoid the attractive pitfall of thinking that we can get something for nothing, that we can get something free.

Energy effectiveness improved

One of the methods of providing for the energy needs of society currently being considered (in fact, it has been employed for more than half a century) is the integrated conversion plant or cogeneration cycle. This type conversion plant conceptually utilizes an energy conversion cascading system, wherein the energy rejected from a second law process is beneficially used in a lower grade form in a thermal process, thus satisfying two needs. With the combination of the proper form of prime movers and in cases where the shaft and thermal loads are adequately coincident and

of the proper magnitudes, such plants can and have been beneficial both from the standpoint of monetary savings and *energy effectiveness (Ee)*. To force adequate coincidence in the loads, the designers of buildings and systems have the opportunity to some extent of controlling the magnitude of the loads and, as a last resort, of providing thermal storage.

However, the whole system concept enters the area of misuse when, to raise the thermal efficiency of the second law process, attempts are made to *build* the thermal load in the normal valleys. This concept of "building the load" often creates a false need for thermal energy under the mistaken notion that the heat energy (generally steam) is free!

Consider economics and *Ee*

Let us consider the economics and *Ee* of a system, considering three simple examples. The conditions are:

• Cost of purchased electricity is 2¢ per kW-hr.

• Heat rate of the utility is 11,377 Btu per kW-hr (*Ee* = 0.30).

• Higher heating value of coal is 11,000 Btu per lb.

• Coal cost for on-site generation is $28 per ton ($1.27 per MMBtu).

• Seasonal boiler efficiency (on-site) is 75 percent.

Case 1

Power is to generated on-site with a steam Rankine cycle, throttle conditions, 400 psig and 750 F; exhaust, 2 in. Hg; 1000 Btu per lb (admittedly low, but adequate for these examples) and 70 percent isentropic efficiency are assumed for the turbine. The heat required to generate 1 kW and the cost is shown in Fig. 43-1. The fuel cost is 1.80¢ per kW-hr and the

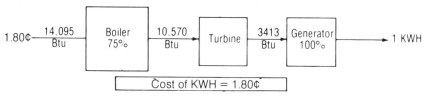

Fig. 43-1. Case 1.

Ee is 24 percent, compared to 2¢ per kW-hr and 30 percent for the utility company.

Case 2

The same energy community has a need for 30,587 Btu thermal energy coincident with the need for 1 kW electrical energy. The conditions and assumptions are the same as in Case 1, except that the turbine will exhaust at 100 psig. The turbine heat rate per kW-hr rises to 34,000 Btu (part (a) of Fig. 43-2) and the plant input fuel requirement is 45,333 Btu, at a cost of 5.75¢. The value of the cascaded heat is represented by the alternative method of generating it with a simplex steam boiler cycle (part (b) of Fig. 43-2), with an indicated value of 5.18¢. The difference, then, is the true cost of the fuel for the kW-hr at 0.57¢. In this case, the *Ee* for the combined plant is 75 percent. With utility electricity and on-site steam, the *Ee* would be 65 percent.

In Case 2, it is quite likely that the appropriate savings of 1.5¢ per kW-hr could amortize the cost of the generating plant and cover such costs as management, operations, main-

tenance, and so on. However, in the vast majority of energy communities, the need for the hypothetical 30,587 Btu of thermal energy is seasonal in nature, and some balance between the favorable Case 2 and the highly marginal Case 1 is strived for by utilizing variable extraction at the 100 psi level. It is at this point, the striving for the advantages of Case 2, that the pitfall in the logic is found, i.e., *building the steam load.*

Case 3

Case 3 is the same as Case 2, except that the community has a requirement for 1.5 ton-hr of cooling coincident with the 1 kW electricity. This is provided as shown in part (a) of Fig. 43-3, with all conditions the same as Case 2, except that the 30,587 Btu is generating the 1.5 ton-hr through an absorption cycle. An obvious alternative to the absorption cycle would be to purchase electricity from the utility and produce the 1.5 ton-hr through a compression cycle consuming 1.5 kW-hr (COP of 3.5) at a cost of 3¢. Subtracting this credit from the 5.75¢ fuel cost produces a net cost for the

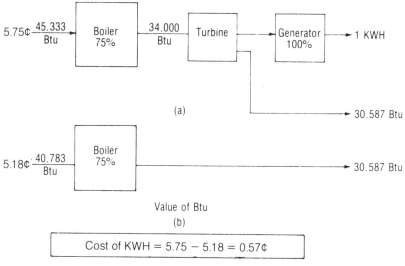

Fig. 43-2. Case 2.

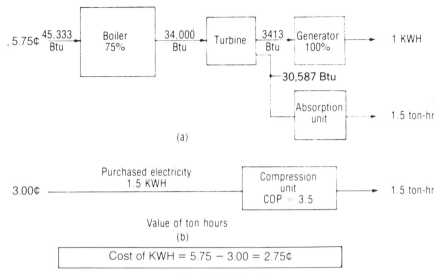

Fig. 43-3. Case 3.

generating kW-hr of 2.75¢. The energy effectiveness for this combined cycle, based on a standard COP of 3.5, is 19 percent, and the total cost is 5.75¢. If the value option were employed—utility power for compression cooling and on-site power for site electricity—the cost would be 4.8¢ and the *Ee* would be 27 percent.

The purpose of this simplified example is to illustrate that so-called free steam is, in fact, not free if the load must be created to provide a use. The decision making relating to combined plants must be based on in-depth analysis of all the dynamic operating conditions relating to the true connected loads and *not* to the loads that are created within the conversion plant itself to increase the so-called efficiency of one specific module or subsystem.

For an economics relationship for this type of conversion system, see Chapter 41.

44

An oil-fired integrated plant design

In 1969, new construction on the Pennsylvania Turnpike at Sideling Hill, bypassing a two-lane tunnel and access to an existing service plaza (gas station and restaurant), necessitated the erection of a new service plaza. This new service facility, approximately 50 percent larger than any other on the Turnpike, consisted of a service station with 62 gasoline pumps and 4 diesel dispensers, parking for 315 vehicles, and a restaurant seating approximately 300 people, with counter, cafeteria, and table service available, in addition to picnic facilities and carryout service.

The service plaza location is approximately 10 miles east of Breezewood, in an area where the only energy utility service available was electric power. Most of the plazas along the Turnpike are relatively self-serving facilities, with their own water and sewage disposal systems; heat is normally supplied by oil-fired furnaces or boilers, summer cooling by electric-driven direct expansion machinery, and most cooking is done with liquefied petroleum gas (propane).

Total energy system reasoning

Following extensive studies, it was decided by Humble Oil Co. to power the Sideling Hill Plaza with an oil total energy system, thus making the facility relatively self-sustaining. The primary reasons for this decision were:

This chapter was reprinted from *ASHRAE Journal*, April 1969 and was originally entitled, "Oil Total Energy Plant Makes Turnpike Service Plaza Self-Sustaining." Appearance of this material in *Energy Engineering and Management for Building Systems* does not necessarily suggest or signify endorsement by the American Society of Heating, Refrigerating or Air-Conditioning Engineers, Inc.

1) The plant could provide electric service reliability not subjected to the vulnerability of long-distance overhead distribution systems.

2) The availability of self-generated power would make the installation of an electric kitchen economically practical, thus eliminating the need for propane storage at the site.

3) Economic studies indicated that the differential investment cost of the central plant (in lieu of commercial electric service and incremental heating and cooling units) could be obtained within the appropriated budget, and would generate income of sufficient magnitude to prove the differential added cost a profitable investment.

4) The plant would provide Humble Oil Co. with a working oil total energy unit, which could be used for purposes of information, education and research, as well as market and sales promotion.

The secondary design criteria were established by the physical relevancies of the facility, and the calculated load profiles of the various energy demands for the project. These were:

1) The plaza was to be open 24 hr per day, and semiskilled, responsible technical personnel were to be on duty at all times. This led to the installation of a semiautomatic rather than a completely automated total energy plant.

2) Because of the remote location, simplicity of machinery and components was an important consideration. This was included in the design formula in this manner: "Wherever possible, consider the use of machinery which can be serviced by locally available technicians rather than highly trained specialists; where

this is not possible, use duplicate apparatus, and employ machinery which has an established reputation for a high degree of reliability."

3) Because of the obvious economies of locating the plant near the major loads (electric and thermal), and the promotional advantage to Humble of having it readily available for visual inspection, the plant was to be located as close as possible to the restaurant building, and preferably at grade level.

4) None of the electric loads were what might be considered frequency critical; i.e., regulation in the range of ± 1 percent would be acceptable.

Design approach

The first step in the design of any on-site power generating plant is to achieve a complete understanding of the physical plant and the projected use of the facility, and from these data develop load profiles for the various forms of energy which the facility will require. Based on these profiles, the prime machinery is selected. The importance of this step is of such magnitude that it cannot be emphasized too strongly. First, if the machinery is of inadequate capacity, the catastrophic results are obvious; and secondly, if the power-producing machinery is oversized, the penalty paid by the enterprise is twofold:

1) The initial cost of the machinery and ancillary equipment increases, thus burdening the plant with an investment cost normally difficult to recover.

2) If reciprocating prime mover equipment is used, the fuel rate (pounds of fuel or value of input heat per unit of mechanical output) increases at an exponential rate, below a power output of approximately 50 percent of maximum continuous rating (Fig. 44-1). Thus, if the prime movers selected are of excessive capacity, the fuel rate will be high and, consequently, the plant will be inefficient and costly to operate. In addition, engine maintenance costs are higher on equipment regularly run under light load conditions.

Due to the unique features in design and use of various building structures, there is little, if any, documented information available on the load profiles of the electrical and thermal (heating, cooling, domestic water heating) requirements of individual buildings. In attempting to predict these requirements for proposed or non-existent facilities, the designer must resort to combining various known factors with educated estimates of unknowns to obtain the composite result. In this case, a facility as similar as was available was used as a model. The basic similarities and differences are shown in Table 44-1. Many obvious differences existed between the facilities, yet the similarities provided a starting point; i.e., a model from which a prototype performance could be developed.

The steps used in projecting information obtainable from the model into reliable data for the prototype were:

1. A detailed inventory of all connected electrical loads at the model facility was made.

2. An estimate was made of all connected

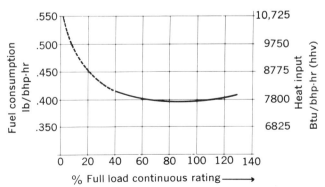

Fig. 44-1. Typical fuel consumption curve turbo-charged and after-cooled, 1200 rpm, four-cycle; diesel engine.

Table 44.1.   Physical Characteristics of Model vs Prototype Facility

	Model facility	Sideling Hill
Restaurant seating	200	300
Property size (acres)	12	37
Paved area (acres)	5.5	9.4
Gasoline dispensers	40	62
Diesel dispensers	2	4
Crossover ramps	No	Yes
Restaurant area (sq ft)	9000	12,000
Kitchen	Propane gas	Electric
Heating system	Oil-fired boilers	Central plant
Cooling system	DX electric	Central plant

electrical loads that were anticipated for the new facility except for cooking apparatus.

3. Recording kw meters were installed at the model facility (one on the restaurant meter and one on the station meter) for typical weeks in November when the electric air conditioning system was not in operation. The sum of the two meters is shown on curve A, Fig. 44-2.

4. The results obtained were upgraded in direct proportion of selective connected load vs. selective connected load (curve B, Fig. 44-2).

5. To the above were added the estimated central plant burden, and the electric kitchen, the kitchen being perhaps the most speculative approximation of the entire load profile development. The predicted kitchen load was developed by tabulating all connected kitchen cooking devices and assigning to each a multiplier (<1) signifying their respective contribution to a coincident load, and summing the products. The end result was to allow approximately 20% of the installed capacity for cooking or kitchen load. The cumulative value of these loads is also indicated on Fig. 44-2.

At this stage in development of plans for the facility, many approximations in lighting loads, motor loads, resistance heating (cooking) equipment had to be necessarily made, since the actual design is yet in the embryo stage.

After the electric load profile was developed, a similar procedure was followed in developing the thermal load profiles. Although over designing or overestimating of these loads was not as

disastrous to the economic feasibility of the plant as a similar error in electric load predictions, such errors could lead to unwise decisions in the selection of recovery or production apparatus, control devices, etc., which could have a detrimental effect on the plant economics.

The development of the thermal load profile is also considerably more complex than the electric, because of the three-dimensional effect of seasonal weather variations; whereas, in most buildings, seasonal fluctuations in conventional electrical profile are easy to predict, once a base profile has been established.

Although no attempt will be made to review the development of the thermal load profile, the elementary ingredients are:

1) Ventilation rates.
2) Predicted building occupancy: nonbusy days, busy days, and average days.
3) Seasonal frequency of each type of occupancy.
4) Building use schedule.
5) Space heating (transmission) loads.
6) Transmission and solar cooling loads.
7) Local weather profile for a typical year, including frequencies, extremes, coincidence of wet and dry bulb temperatures, wind velocities, and cloud cover.

From the above, were constructed three daily thermal (heating and cooling) load profiles for each month. These profiles are compared to the electrical load profile (which, by changing scale, becomes a salvage heat availability curve), to determine both the demands and predicted consumptions of salvage heat availability.

This study indicated that for this particular project, the utilization factor of the heat recovered from exhaust gases would be too low to justify the initial cost investment of heat recovery mufflers. Thus, it was decided to allow this potentially available portion of the input heat to be wasted.

Prime mover selection

Figure 44-2 shows that the minimum anticipated electrical load is approximately 65 percent of the maximum. Reference to Fig. 44-1 indicates that a diesel engine prime mover will

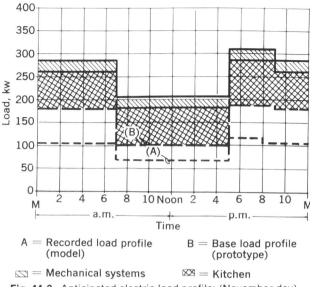

A = Recorded load profile   B = Base load profile
(model)   (prototype)

⟋ = Mechanical systems   ⧓ = Kitchen

Fig. 44-2. Anticipated electric load profile; (November day).

operate in the range of optimum efficiency down to 50 percent of the full load continuous rating. The value of this comparison is that a single prime mover could be selected which in all ranges of anticipated loads would be on the optimum part of the fuel rate or efficiency curve. (In plants where the minimum loads are less than 50 percent of maximum, three or more engines are often employed to achieve fuel economy.) The shaded portion of Fig. 44-3 indicates the range of optimum efficiency of the units selected for this project: Line A represents the full load continuous rating, Line B the anticipated load profile. It can be seen that the unit is of adequate capacity to handle the maximum anticipated load with a little reserve, and the fuel rate does not vary by more than 2.5 percent throughout the entire range of expected use.

The final decision, therefore, was to provide two units of identical capacity, one unit operating at any given time, with the second unit available as a standby. The electrical power and control systems were arranged such that

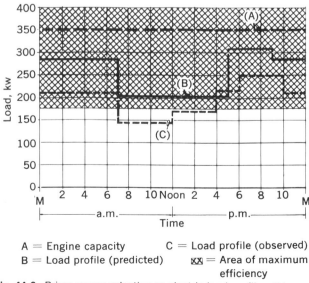

A = Engine capacity   C = Load profile (observed)
B = Load profile (predicted)   ⧓ = Area of maximum
                                      efficiency

Fig. 44-3. Prime mover selection vs electric load profiles; (November).

in the event of a failure of the "on" unit, the following sequence occurred:

1) If the unit fails, low voltage causes undervoltage trips to drop out all nonessential circuit breakers.

2) A dead bus relay initiates the cranking cycle for the second or standby unit.

3) When the standby unit establishes speed and voltage, the generator breaker closes and picks up all essential loads (those which are not equipped with undervoltage trips).

4) One of the assigned personnel from the station closes the remaining breakers manually, checks failure lights to determine cause of shutdown, and notifies the service supervisor by phone.

This sequence requires approximately 8 to 12 sec through the first three stages, and all loads should be reestablished in not more than a few minutes.

Since one of the original design criteria was that there were no frequency-sensitive loads, the units were provided with high-quality (1 percent regulation) hydraulic governors in lieu of more highly sophisticated electronic sensing and/or controlling units. The hydraulic units are rugged, time-proven devices, relatively inexpensive, which can be adjusted by the attendant maintenance personnel, and provide frequency regulation within the limits required. A simple comparator clock system was provided for periodic fine tuning or adjustments of speed. Since the plant is a "one-unit plant," the speed droop problem normally associated with stable operation of parallel hydraulic governed units was not relevant. The only time the units are operated in parallel is when the selected "on" unit is being alternated.

Heat recovery system

The potential fuel energy entering a supercharged diesel engine generator unit is converted into mechanical energy and heat (assuming no carbon in the exhaust). The mechanical energy is converted to useful shaft work which is used to drive the generator, and in turn produce electrical energy, and to overcome friction and windage. The units

selected for this project were supercharged, high-compression ratio, 1200 rpm, with a thermal efficiency of 29 percent (electric output versus heat input) at full load. In addition to the electrical output, heat energy from both thermodynamic and mechanical inefficiencies is ultimately found in the following forms of dissipation, in the approximate amounts indicated.

Electrical power	29 percent
Engine jacket coolant	21 percent
Aftercooler	4 percent
Oil cooler	10 percent
Nonrecoverable exhaust	10 percent
Recoverable exhaust	15 percent
Radiation and convection	8 percent
Unaccounted for	3 percent
Total = Input	100 percent

Figure 44-4 shows the above heat balance in graphical form, and also incorporates the method employed in the plant for utilizing or dissipating each form. Although it is quite common in many total energy plants to use the recoverable exhaust heat, the oil cooler heat, and the aftercooler heat, thus producing a combined plant efficiency of approximately 80 percent, this combined power and recovered heat only has value if the coincident load profiles indicate a relatively high-utilization factor for the available heat. Storage schemes can be applied to control this coincidence to some extent, if the space usage is regular and readily anticipated (such as in a school or office building). However, load profile comparisons on the Sideling Hill project indicated that although 100 percent of the potentially available heat could be used at times, the actual hours per year of much of this use would be quite low. An additional consideration in the selection of heat recovery systems for diesel engines of the type employed on this project is that if the jacket coolant systems are operated as high as 240 F (approximate saturation temperature of 10 psig steam), the engine lube oil must be cooled to approximately 180 F entering the engine, thus making recoverable heat at temperatures in excess of 140 to 160 F impractical.

On the basis of the foregoing, the decision

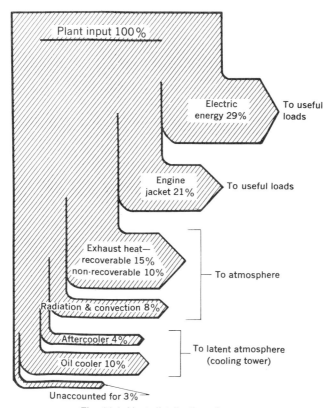

Fig. 44.4 Heat distribution diagram.

was to recover only the engine jacket heat, which, when combined with the electrical output, would provide a plant of 50 percent combined efficiency, and virtually 100 percent utilization of this 50 percent of the input energy potential.

An ebullient cooling system for the engine jacket was selected, since, in keeping with the original design criteria, optimum reliability could be obtained at minimum cost. Each engine was provided with its own steam separator, flow controls, water level controls, safety controls, and steam pressure controls. With this arrangement, a pipe rupture, or any other malfunction in the jacket coolant system of one unit would have no effect on the other. Another advantage of the ebullient system is that the circuit depends upon no mechanical device to create or assist circulation of the coolant.

Steam pressure in the steam separators is controlled by a back pressure regulating valve between the separator and the steam header,

holding a minimum pressure of 10 psig on the separator. Thus, the engine jackets (after warm-up) operate at a constant temperature of approximately 240 F. The header pressure is controlled between 5 and 14 psig. A pressure switch in the header turns the boiler on (Fig. 44-5) at 4 psig and off at 8 psig. If the header pressure should exceed 14 psig, a pressure control valve passes steam to a waste heat condenser which rejects the heat through the cooling towers. Note that this will happen only when the heat supply from the salvage system exceeds the demand.

Heat utilization systems

The flow diagram of Fig. 44-5 shows that steam recovered from the engine jacket is piped to two places in addition to the waste heat condenser: the absorption refrigeration unit and secondary circuit water heat exchangers. The use in the absorption unit is obvious; i.e., steam is condensed in proportion to the restaurant cooling load (peak load

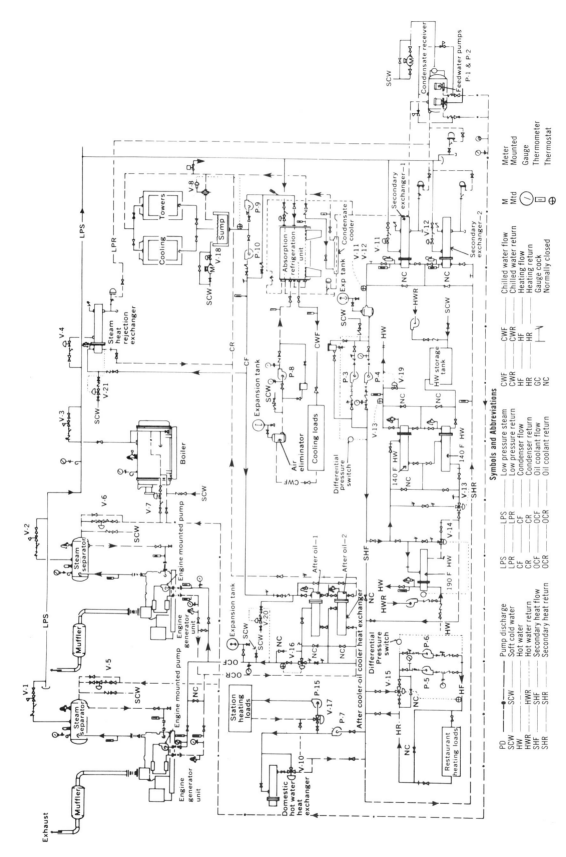

Symbols and Abbreviations

PD	——●——	Pump discharge		LPS	———	Low pressure steam		CWF	—CWF—	Chilled water flow
SCW	—SCW—	Soft cold water		LPR	———	Low pressure return		CWR	—CWR—	Chilled water return
HW	———	Hot water		CF	———	Condenser flow		HF	—HF—	Heating flow
HWR	—HWR—	Hot water return		CR	———	Condenser return		HR	—HR—	Heating return
SHF	—SHF—	Secondary heat flow		OCF	———	Oil coolant flow		GC		Gauge cock
SHR	—SHR—	Secondary heat return		OCR	———	Oil coolant return		NC		Normally closed

M	Mtd	Meter Mounted
		Gauge
		Thermometer
		Thermostat

176

is 60 ton or approximately 1200 lb of steam/hr).

The heat which is transferred to the secondary circuit takes a less direct route to its point of ultimate consumption. Since steam pressures in excess of 5 psig (227 F) would not be available at all times, the secondary heat circuit was designed to operate at a maximum water temperature of 210 F. The six-valve bypass around the exchangers coupled with the "either-or" pump arrangement is, again, designed for optimum reliability. Although both exchangers are required during the severest winter months, either one will suffice in summer or midseason weather, allowing the maintenance staff ample time for heat exchanger maintenance, and providing at least partial heat availability in the event of a component failure. The secondary pumps are sized and arranged such that only one pump handles the circuit requirements: the same is true of the restaurant heating pumps. If the operating pump should for some reason fail, a pressure-actuated switch in the circuit would turn on the standby pump.

Since the maximum design temperature in the secondary circuit is only 210 F, it was necessary to arrange the system in a series utilization scheme to assure that, even under severest demand conditions, ample temperatures were always available where needed (with minimum flow rates for economic reasons). The establishment of design temperatures for the hot water reheat and radiation systems was, within relatively wide limitations, up to the designer's selection; whereas the necessity of 190 F domestic hot water for the dishwashers was a mandatory requirement. Thus, a study of Fig. 44-5 reveals that the secondary water is used in series, first for the 190 and 140 F domestic water heating systems, next for the restaurant heating circuit, then finally for the service station heating sytem. All exchangers and heating apparatus are designed to cope with this arrangement of reduced temperatures along the circuit, and the total design temperature drop is 40 F deg.

If the engine jacket coolant circuit had been selected with a lower temperature cooling scheme, and as a result lube oil coolant as high as 180 F had been used, the logical point for connecting this source into the system would have been the secondary heat return. This would have facilitated the salvage of an additional 10 percent of the heat input. However, as previously stated, it was decided that the advantages of maximum reliability at minimal investment led to the use of 10 psig steam for jacket cooling, and that this in turn prevented the possibility of using lube oil heat, due to the low coolant temperature required.

Heat dissipation systems

As stated previously, energy supplied by the fuel is converted to shaft work and heat. The shaft work is utilized in the ultimate form of electrical energy, and that portion of the heat extracted from the engine jackets is utilized as described above.

Reference to Fig. 44-4 indicates that heat is also found in the aftercooler, lube oil cooler, exhaust, and radiation and convection from the prime movers and generators. This heat, of course, must be dissipated to the atmosphere or some other available heat sink. Furthermore, the reliability of the dissipating systems is as critical as that of the reclaiming systems, since any failure in the plant ability to dissipate the heat in any available form will ultimately result in a complete plant shutdown.

In a supercharged engine of the type used, an exhaust gas turbine drives a centrifugal compressor in the intake combustion airstream. The air is compressed to approximately 2 atm (30 in. Hg gauge), a process which can be assumed to be isentropic, and in which the air temperature is raised by the relationship:

$$T_2 = T_1 \left(\frac{P_2}{P_1} \right)^{\frac{k-1}{k}}.$$

Substitution of the appropriate numerical values indicates that if the entering (ambient) air is 90 F, the temperature leaving the compressor is 214 F. This air must be cooled to a lower temperature, or the mass entering the cylinder (due to high specific volume) would be quite low, and the advantages of the supercharging would be lost. The heat removal required to cool the air is:

$$Q_{ac} = M_a(h_2 - h_3)$$

or

$$Q_{ac} = M_a C_p(T_2 - T_3)$$

where

Q_{ac} = heat removed at aftercooler Btu/min,

M_a = mass flow rate of combustion air lb/min,

h_2 = enthalpy of air entering aftercooler,

h_3 = enthalpy of air leaving aftercooler,

C_p = average specific heat or air (at constant pressure),

T_2 = temperature of air entering aftercooler,

T_3 = temperature of air leaving aftercooler.

(This formula assumes the kinetic energy effects $(V^2/2_g)$ of the entering and leaving air are equal.)

The temperature of the air leaving the aftercooler has a direct relationship to the prime mover capacity. As shown earlier, it was anticipated that full load capacity may be needed at times. Proper design required that the lowest temperature cooling medium available be used for the aftercooler cooling circuit.

Since well water at the site is available in limited quantities, and extremely deep (therefore costly to pump), cooling towers were selected as the heat sink. Tower water is pumped from the underground reservoir (pump P-10) to a pair of heat exchangers (called after-oil exchangers). The reason for this stage of heat exchange is that it was considered beneficial to have readily cleanable exchangers for all open water circuits. Again, as in the case of the steam-to-secondary exchangers, two units are provided, with six-valve bypasses on both sides, providing an either-or both selection to facilitate maintenance. The circuit on the shell side of the exchangers is operated at a control temperature of 110 F and water pumped by engine-driven pumps is circulated in series first through the aftercooler, then through the oil cooler.

The steam rejection heat exchanger is in parallel with the after-oil heat exchangers. Water from pump P-10 is available to this unit at all times and, when the steam header pressure exceeds 14 psig (less requirement for heat than is available from the jacket), the pressure controller opens the steam valve.

Due to the critical nature of the heat rejection circuits, multiple backup devices were employed with little increase in initial cost. In the event the rejection pump, P-10, fails, the absorption unit condenser water pump, P-9, can be used to circulate through this circuit by proper arrangement of the manual valves. In the event both pumps are inoperative, or the rejection pump becomes inoperative and the condenser pump is not yet put into service for rejection, self-contained temperature-actuated valves (V-20 and V-21) will open and feed soft cold water through the two exchangers and to the sump until the problem is remedied.

Since exhaust heat is not used in any form, this heat is readily dissipated directly to the atmosphere, after the gases are passed through a residential quality muffling system.

The radiation and convection heat from the engine and generator is dissipated directly into the room. Thus, the only direct cooling or removal medium available is the room air, and the least costly method of ultimate removal is by circulation of adequate quantities of outdoor air through the space. The quantity of ventilation air required is readily calculated by the sensible heat relationship:

$$q_{re} = \text{cfm } (1.10) (t_{in} - t_{out}).$$

Substitution of values for a 350-kW unit, with a room temperature rise of 10 F indicates a required ventilation rate of 30,000 cfm.

Since this is an unreasonably high volume of air to move through a limited space, the most logical approach is to accept temperature increases in the order of 30 F, and carefully control the flow pattern. The method employed on this project was to introduce air from a nominal 10,000 cfm supply fan, through a system of louvers located just beyond the generator end of the assembly, discharging the air vertically downward so the generator and engine combustion air intakes receive the coolest available air. The airflow pattern moves across the engines and mufflers and discharges through an acoustically treated, directional roof-mounted relief outlet. The supply air dampers are automatically controlled so that only the dampers directed at the "on" unit open, and a room thermostat regulates the damper position to prevent overcooling of the room during cold weather.

Other systems

In any total self-dependent-type facility, all subsystems are of the utmost importance. Every item required for reliable operation, regardless of how small or seemingly unimportant, is much like one of the thousands of small links in a long chain—if it fails, the entire chain is worthless. The inclusion of this realization into the design formula is imperative if the plant is to achieve the reliability anticipated. Careful attention must be given to such seemingly insignificant items as details of construction of water level controllers and safety cutouts, reliability of electrical relays, quality of seals, piping joints, range of accuracy of pressure-relief valves, and so forth; any of which could, in the event of malfunction, cause a plant failure. There are different methods employed in achieving this reliability in some of the other major systems.

The fuel system consists of an underground storage system with a capacity of approximately 30 days, which can receive fuel in full truckload quantities (for minimum cost). From the storage system, the fuel is pumped by a fuel pumping unit into a day tank which holds, below the low alarm level, 4 hr supply. Two transfer pumps are operated by a manual selection of "either-or." The selected pump transfers the fuel by command of a signal from a high-quality float control on the day tank. The control is arranged so that if the oil reaches a level below that at which the selected pump is supposed to turn on, the standby or second pump will take over. If this second pump fails, or if for any other reason the level drops below a predetermined point lower than the second pump-on point, the low-fuel alarm sounds, alerting the station attendant that within 4 hr, the plant may shut down.

The temperature control system for the plant is pneumatically operated, except for the two backup controls incorporated into the heat dissipation systems (V-20 and V-21). This compressed air system is also used for engine cranking power. Thus, the compressed air may be considered as one of the important stored energy sources in the plant. Two electrically driven compressors are employed, controlled in an "either-or" arrangement, such that the selected unit is the slave. A two-stage pneumatic-electric switch turns on the backup unit if the selected unit fails to operate or cannot keep up with the load (a condition resulting from a system leak). One of the base-mounted motor compressor units is provided with an auxiliary gasoline engine drive, which is manually linked to the compressor by a belt reversal arrangement, and manually started. This system required for initial start-up is available for any subsequent dead start that may be necessary.

The air from the storage system (at 90 to 140 psig) is piped directly to the engine starter regulators. Temperature control air is first passed through a dryer unit, then reduced in pressure to approximately 25 psig, and routed to the temperature control systems. All components in the temperature control systems are arranged for a predetermined fail-safe mode of operation.

Engine-generator unit controls, both operating and safety, are energized by a 24-V dc source primarily served from a system of four 12-V heavy duty, lead acid batteries, connected as two series circuits in parallel. Even complete failure of one battery or one connection would still leave the plant with a 24-V source. This series-parallel storage battery system, during normal operation, floats on the dc system which is energized by a rectifier circuit from the 110-V ac source which, in addition to actual operation of the control system, continually maintains a controlled charge on the four batteries.

Electrical system

The voltages required for terminal apparatus were established on the same basis as would be employed in any other project. Electric cooking apparatus was selected at 208 V because new as well as replacement components are readily available at this voltage. Yard and ramp lighting was selected at 265 V because the fixtures were standard, and provided a highly efficient distribution system. Building lighting was selected at 120 V since a large percentage was incandescent. Selection of a voltage for three-phase motors was delayed until the generation voltage was determined.

Comparison of the voltages required led to the selection of generating at 265/460 4-wire Wye. The generator output voltage was then used directly for primary distribution, yard and ramp lighting, and plant motor loads. It was transformed locally through dry-type transformers to 120 and 208 V, as required for lighting, convenience outlets, and kitchen apparatus.

As stated earlier, if the generating unit fails, a dead bus relay initiates the cranking cycle for the standby unit. To prevent an overload of the oncoming unit, the circuit breakers for all nonpreferred electrical loads were provided with undervoltage trips which require manual resetting following a failure.

Conclusion

A considerable number of other design decisions were required in the development of the plant. However, limited space restricts the inclusion of a discussion of all these decisions. This omission does not in any way lessen the importance of these links; including space arrangements, control panel arrangement for operational ease and simplicity, isolation of vibration and noise, etc.

The plant, following several weeks of intermittent operation, a period used for shakedown or debugging, was put on the line for permanent operation November 26, 1968. The performance has been as desired and anticipated. Although original designs made pro-

visions for instrumentation at a later date, none (except for operational, accounting, and maintenance records purposes) have been installed. However, in an effort to develop the actual load profile, hourly demand readings were taken one late November day, and the results are shown as Curve C, Fig. 44-3. Note that the general shape of the curve is similar to the predicted curve, except that the magnitude is somewhat less. Since this was only one day's readings, no reliable conclusions can be drawn, but a procedure for assimilating additional data regarding load profiles and space usage description for ultimate comparison against the early approximations has been initiated. With this information, and hopefully, the installation of a recording wattmeter in the near future, any deviations between the two profiles will be understood.

Again referring to Fig. 44-3, since the actual profile is lower than the predicted, note the effect this deviation would have on the plant economics if further data were to confirm this as a typical day. Study of the available machinery selection data indicates that the next size smaller unit could not have been used, since it would have been of inadequate capacity to handle the peaks realized during the cooling season. It is likely engine selection would have been the same had the actual profile been predicted accurately. From the standpoint of operating economy, between the hours of 7:00 A.M. and noon on the day recorded, the unit, because of low load, consumed approximately $4\frac{1}{2}$ percent more fuel than if it had operated as predicted. Again, assuming that this profile was representative of all noncooling months, this increase in fuel rate would prove to be negligible on an annual consumption basis (less than 100 gal/yr).

The plant experience in other respects, has been very much as anticipated, from the standpoints of performance, reliability, and economics.

45

The status of total energy in 1980

Total energy systems (more recently reborn under the name of cogeneration) was a term coined to encompass energy conversion systems for building projects that basically provided, from a fossil fuel source, all of the end-use energy forms required for the project. The concept is philosophically and technically sound from the standpoint of increasing the effectiveness of the use of energy resources. Simply stated, the energy rejected from the power cycle is held at a high enough temperature level to enable its use to satisfy the thermal energy needs.

With properly balanced loads, if this conceptual efficiency is achieved, the cost of energy to the building owner is reduced. Furthermore, the fuel supplier has a beneficial customer, in that the annual load factor is usually much higher than for, say, a heating only or cooling only plant.

These mutual advantages for owner and fuel supplier converged, catalyzed by some inventive designers and state of the art availability of hardware in the early 1960s. For a period of years around 1965, many such plants were designed and installed in commercial and institutional buildings throughout the United States.

Ten to fifteen years later, many such plants are operating well, in the manner initially intended. Many others, however, have been or are being removed, to be replaced by other types of conversion systems or energy source forms. The question obviously might be asked in light of the failure of numerous plants, what went wrong? Was there something wrong with the concept? Has the aftermath of the so-called energy crisis unfavorably changed the economics of source energy?

The answer is not simple, but there is nothing wrong with the concept. Nor are source energy economics the culprit. There are several reasons for the failure of the plants, any one of which could be a fundamental cause for any specific plant, but in most cases there were combinations of causes prevalent.

High level of skill required

The first cause is failure of the building owning-managing team to recognize the management responsibility associated with having such a plant. To manage and operate an integrated conversion plant requires a relatively high level of technical and management skills. In many case histories, when this problem became apparent, the owners found themselves contracting out this responsibility to another agency. Two problems resulted. First, since the management requirements had not been identified initially, the costs associated therewith had not been provided for in the pro forma. Thus, the economics of the plant changed significantly. Second, the agencies contracting to undertake the management of the plants, in many cases, were themselves incapable of providing the technical expertise required. As a result of this, not only did the cost of operations increase, but the anticipated performance was not achieved.

Other causes of failures

The second cause of failure was that the designers of many such plants were embarking upon new technologies with which they were not familiar. These designers came from the ranks of thermal building systems experience. The integration of prime movers, heat recovery apparatus, electrical generation apparatus, and the like introduced elements of major hardware with which they were not familiar. In some cases, basic hardware was misapplied; in others, due to insufficient historical experience in the industry as a whole, the sizes of the conversion modules were improperly selected. Problems resulting from these deficiencies in design were premature failure of machinery, catastrophic interruptions in service, higher than necessary investment costs, and higher than anticipated fuel costs.

The third cause of failure touches a serious misunderstanding of the fundamental concept—that is, a failure to recognize both the static and the time-integrated nature of the loads as they relate to the conversion system. This lack of understanding led to the installation of many integrated plants that were doomed to failure. The building designer has a significant element of control over the magnitude of both the thermal and electrical design loads. He has the opportunity in the conceptual stage of design not only to minimize these loads but also to design to achieve a balance compatible with that provided by a given power-thermal cycle design. The error that often was made was that of trying to force this balance in design of the *plant*. Such forcing simply led to the creation of false loads that improved the thermal efficiency of the combined cycle machinery, while increasing the cost of fuel consumed when compared to alternative methods of providing energy to serve the same real building loads.

As a result, many plants have achieved extremely high power-thermal cycle efficiencies while being economically disastrous.

Integrated energy systems

Now, at a time when most sectors of our society have matured in their recognition of the need to utilize our energy resources most effectively, we would benefit by considering the advantages of integrated energy conversion systems. We *must*, however, learn from out past experience regarding total energy. The U.S. Department of Energy is expending extensive efforts in the direction of integrated energy systems; we should hope they will address the historical experience. If they do not, these efforts will be destined for the same fate as the total energy boom of the sixties.

It is alarming to observe that there is one more reason for the abandonment of total energy plants. That is, allocation and curtailment of fuels. In case history examples, integrated plants that were both energy efficient and economically successful were abandoned because the fuel source was curtailed! This is one more example of the need for a technically mature national energy policy that acknowledges the efficient use of energy resources where they exist.

SECTION IX

Liquid and two-phase thermal fluid systems (hydronics, steam, and refrigerants)

Liquid and two-phase thermal fluid systems are fundamentally energy transport systems. Their basic role in the overall building environmental system is to receive energy from the available source in the thermal form, and by undergoing either a sensible or latent change, convey it to some point of use, and there dissipate it for whatever intended purpose. The reader may naturally assume that the intended purpose is to heat or cool the space, but this is neither the only use nor the most likely. Other uses of thermal fluids include driving power machinery such as turbines and steam engines, motivating absorption refrigeration, and control heat (such as is required in reheat and dual stream systems).

Most textbooks and other publications classify the materials of this chapter as "steam, hot water, and chilled water systems." The reason for reaching toward a more fundamental classification of this grouping of subsystems relates to an extremely important concept. That is, that many areas of technology cease advancing because of artificial boundries drawn about them for the sole purpose of identification. These boundaries tend to place a constraint upon the practitioners in the respective fields. One example of this observation is the chapter entitled "Vapor Lock in Refrigeration Systems." Although this phenomenon had not (to the best of the author's knowledge) been published prior to the publication of this original article, it is quite a common consideration in hydronic system design.

Steam heating systems serve as the classical example of this observation.

When technology in single-phase fluid systems (hydronics) advanced to the stage that their use solved the load control shortcomings of steam radiation systems, steam was replaced by water as the primary radiation-type heating system in commercial and institutional buildings. As a result, all development in two-phase heat transport technology apparently ceased. Yet the advantages of the latent heat exchange and transport concept remain to be developed.

There are numerous concepts in load control of two-phase systems that should be researched. Such research, for example, could provide a method (or methods) for retrofitting existing steam radiation systems for energy conservation. The simple reference of two-phase systems as "steam" systems tends to outrule other fluids that might be used in two-phase systems. One use of such fluids other than steam is in solar heat collection systems. There are major problems with single-phase systems in solar collection that might be avoided by the use of two-phase systems. Two such problems are those of freezing and reverse losses. The three most common solutions to the freeze prevention problem are: (1) install the system in a climate where the temperature never drops below 32 F, (2) drain the outdoor system down when the outdoor temperature drops below 32 F *and* the solar heat does not offset the temperature depression, and (3) add an ethelene glycol (antifreeze) solution to the water. All three of these solutions present either problems that result in significantly increased cost or limit the practicability of solar source heat systems. If, on the other hand, a commercially available refrigerant were used to transfer the heat from the outdoor solar collector to an indoor load device the "collection" problem would be greatly simplified, and it is likely that the cost could be significantly reduced.

There are numerous other examples of the opportunities in expanding the technology of heat transfer fluid systems if the artificial constraints of "steam and water" are removed. The chapter "Preheating Outdoor Air with Transfer Fluid Systems" is another example of some ideas that might be developed.

The chapter entitled "Hydronic Systems Overview" includes no new or imaginative technical information; it was written with the intention of presenting the known state of the art in a new perspective. The concept of the hydraulic analysis and thermal analysis has been used as an instructional tool, and once considered it tends to reveal many new avenues of understanding in the technology of hydronic systems.

There are two chapters on the "Integrated Decentralized Chilled Water Systems," one on the pure concept and the other a case history discussion of a system that was planned at the time the chapter first appeared as a magazine article. Since that time the system has been installed, and the performance has exceeded that anticipated. This type of system does not account for a very large share of the larger campus type of systems as of this date. However, with increasing costs of investment funds and energy, and a populace tending toward an acceptance of higher summer dew point temperatures, this concept could well move into a position of being a common alternative to the central plant. (Some current development work is being done on decentralizing but integrating heating systems in a somewhat similar manner.)

The chapter on steam, like that on hydronics, is simply a new perspective of an existing technology. Since this material was initially published, many of the concepts presented in this chapter have been investigated in the laboratory and in actual field installations. As a result, it appears that the old technology of "steam" has gotten a slight nudge forward. Given the inertia that this field appears to possess, this may be enough to get it rolling again.

To conclude, the intermediate fluid systems have been the bulwark of the contemporary large heating and cooling systems. The development of their technology was rapid and almost spontaneous. Now, in retrospect we might consider if we have derived the most out of our thermal fluid technology, or if there is room (or, indeed, need) for significant improvement. If we recognize that this need for improvement exists, this recognition, of itself, will motivate the advances in the state of the art.

46
Hydronic systems overview

Hydronics, for the purpose of this chapter, will be defined as single-phase (liquid) energy transport systems, such as chilled or hot water systems. Over the past two decades, these systems have moved into the position of being the most commonly used heat transport and intermediate fluid energy transfer systems. If the reason for the shift to hydronics could be summed up in a single phrase, it would be *inherent ease of control*.

Although hydronic systems may lack some of the advantages of two-phase systems or all-air systems, their adaptability to multiple zones of load control with highly effective performance results has proved an overwhelming benefit. This benefit has had some backlash, in that the systems could be extremely forgiving of design or installation errors, such as oversizing of loads, sources, pumps, piping, etc., and the inherent control simplicity would correct for the deviations and still provide acceptable performance results.

Fundamentals re-evaluated

Like most new and rapidly growing technologies, the development of hydronic systems "happened" more than it was planned; i.e., much of the hardware development and system design evolution was to address prior problems. Three marketplace pressures are currently present that are forcing a re-evaluation of the fundamental concepts of hydronic systems. They are:

• Awareness of energy economics has resulted in a consideration of the inherent process energy waste resulting from both overdesign and designs that do not take into account auxiliary systems and false loading burdens.
• The spiraling costs of construction and interest rates on investment money have exerted pressures upon the engineering profession to develop systems that can be installed at a minimum construction cost while still satisfying the other performance and design parameters.
• The relatively simple concepts employed in earlier, smaller systems were not reevaluated as the hydronic systems grew in magnitude and complexity. As a result, hydronic systems of a complexity that is virtually impossible to understand have been installed spanning large campus complexes and large, densely populated areas.

The last of these was recognized prior to the energy and cost implications, and its presence or recognition formed the basis for reevaluating the concepts. There are two fundamental aspects to the analysis of a hydronic system that must be recognized if a systems analysis is to be performed: they are the *hydraulic analysis* and the *thermal analysis*. Although these are separate phenomena, they are intimately interdependent.

Hydronic system is hydraulic

The vast majority of problems revealed by the large complex systems was due to the lack of recognition of the hydraulic phenomena. Simply, a hydronic system is a hydraulic system containing a noncompressible fluid. As such, any change in pressure or flow rate in one part of the system, no matter how small or remote, will affect the pressure or flow rate in *all other parts of the system*. The only element of compressibility is the compression tank, which has a fundamental role in the hydraulic analysis. The compression tank contains the liquid of the system and a compressible gas (with either a free interface or separated by a diaphragm). Its salient hydraulic function is to establish the hydraulic constant pressure point in the system. Under operational dynamics, as valves change positions, pumps cycle on and off, and so on, the pressures at *all* other points will change, but at the point of the connection of the compression tank, it will be constant. This is analogous to the electrical concept of ground potential. Except for the ground in an electrical distribution and utilization system, all potentials are simply relative to one another, and an analysis of such a system is impossible to undertake without the ground reference. Similarly, in a hydronic system, any efforts at a hydraulic analysis cannot be undertaken effectively without establishing the ground, which in the analogy is the pressure at the compression tank.

Consider these basic rules

This establishes the first cardinal rule in the hydraulic systems analysis: *no hydronic system, no matter how large or complex, should have more than one compression tank connection.* Multiple tanks can be used if they are piped to function as one vessel and connected to the main piping at a single point.

The second basic component of the system that must be addressed in the hydraulic analysis is the load. The load is the component that transfers thermal energy between the system and the conditioned space or the psychrometric system that conditions the space. Considering the load control from the standpoint of the hydronic system, the load is controlled (reduced from design quantities) by reducing the log mean temperature difference between the hydronic fluid and the air. This is accomplished in most systems by either reducing the flow rate of the hydronic fluid or by reducing the temperature difference between the entering fluid and the entering air (reducing the EWT in heating systems or increasing the EWT in cooling systems). Although the heat transfer rate (or load control) is a thermal analysis phenomenon, if it is accomplished by either of those two means, it affects the hydraulic analysis.

Load control is important

The method of load control has been observed to be one of the most important and least

understood design requirements of the hydronic system. The decision as to what method of load control to apply must be made with careful consideration of its impact on both the thermal and hydraulic characteristics of the system.

From the thermal standpoint, the three methods of control are:

- constant flow and constant entering temperature and variable air side mass flow;
- constant entering or leaving fluid temperature and variable flow in the load circuit;
- constant load circuit flow and variable entering fluid temperature.

The first of these has historically been employed in smaller loads, such as unit heaters, with cycling of the fan or blowers, and in larger systems, with face and bypass control as for preheat coils or hot deck and cold deck coils of multizone or double duct systems. This method has no impact on the hydraulic system; i.e., the hydraulic system does not "know" whether the load at any given time is 100 percent or 10 percent of design. From an energy standpoint, the hydraulic system energy input is always at design quantity. Thermally, because of numerous phenomena, such as damper leakage, excess dehumidification, and so on, the wild flow-coil systems generally impose increased loads upon both chiller and boiler systems, relative to the reductions in space load. Thus, in the interest of energy economics, the use of wild flow systems is generally considered to be undesirable.

The second method, constant entering or leaving fluid temperature and variable flow in the load circuit, has historically been the most common control mode employed. In the simplest configuration, the on-off cycling of the circulator or pump is a form of this mode of control. As the load senses it, this is a frequency modulation or on-off control, and it has generally been limited to smaller heating-only systems. The understanding of the hydraulic impact of the on-off method is fundamentally simple if the constant pressure point has been properly established. If this has not been done, pump cycling can have catastrophic effects.

Valve control is one method

In larger systems, valve control is used to provide the variable load flow. Identical thermal response relating to the load is achieved with either a three-way or two-way throttling valve. The three-way valve is usually a mixing valve, installed at the outlet of the load device so that, as it mixes inlet flow streams, the flow from the bypass stream essentially accomplishes a reduction in the flow rate through the load. The same reduction in flow through the load can be accomplished with a throttling or two-way valve that is installed in the load circuit either upstream or downstream from the load device and responds to load reductions by simply throttling the flow. The selection of the type of valve is predicated upon a system feature other than load control. The differences are fundamentally hydraulic. When a three-way valve is employed, the total flow rate through the system, or at the least load circuit, is essentially constant. This, then, is theoretically a constant flow, variable temperature differential circuit as the hydraulic system sees it and as the thermal source system sees it. Being constant flow, the reduced load energy consumption by the hydraulic system is equal to the full load energy rate. From the standpoint of energy economics, this is an undesirable feature.

The two-way valve, on the other hand, has the feature of impacting the thermal source system with a reduced load characteristic of reduced flow and essentially constant temperature differential. It relates to the hydraulic system by reducing the flow rate and increasing pressure differentials across both the pump and the load circuits. Because of those pressure variations, application of two-way valves requires careful design attention to:

- location of the constant pressure point;
- maximum allowable pressure differential across the valves; and
- characteristic curves of the pumps.

The feature of variable flow reflected on the source generally improves the control stability and simplicity of the source system, but again imposes the requirement of careful design attention to the source apparatus configuration. Because of these crucial requirements of design in "two-way" valve systems, extensive use has been made of the three-way valve. However, from the standpoint of energy economics, this has been a costly alternative. The benefits of reduced energy consumption at reduced flow can no longer be overlooked, and future systems designers and manufacturers will be led to address the design concepts associated with the throttling valve load control systems.

Third control method cited

The third load control method, constant flow and variable entering fluid temperature, is also seen in numerous configurations. In its simplest form, this control method is achieved by varying the temperature of the fluid at the source system. It is commonly called reset control in heating water systems, used where the need for heat varies with outdoor temperature, such as in perimeter radiation or homogeneously loaded dual stream fan systems. This approach provides for stability of control in the case where it is compounded with fluid control, or pure simplicity where it is employed with wild flow circuits. The only hydraulic phenomenon interaction is that the wide range of average system temperatures must be considered in sizing the compression tank.

In larger systems with essentially nonhomogeneous load requirements, this type of control is accomplished by unique load-assigned pumping, commonly called secondary pumping.

Secondary pumping increasing

Secondary pumping is finding increasing use in large, complex systems for the fundamental reason that it provides a *modular* aspect to the design, seemingly simplifying the understanding of these extended systems. However, when improperly applied, secondary pumping can add a degree of complexity and control instability that is impossible to cope with. This is a result of a lack of proper understanding of the hydraulic phenomena.

In its simplest form, the cardinal rule of design for all secondary pumping systems is that the primary circuit must have no dynamic

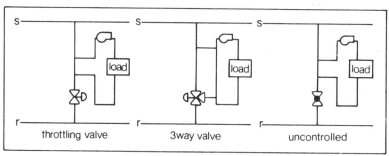

Fig. 46-1. Three basic secondary pumping connections.

hydraulic effect upon the secondary circuit and the secondary circuit must have no dynamic hydraulic effect on the primary circuit. This is a simplified statement of the "common pipe" concept originally published by G. F. Carlson.[1] Although there are numerous configurations of the secondary pumping connections, they can virtually all be reduced hydraulically to one of the three diagrams shown in Fig. 46-1. A cursory study of the three diagrams will reveal some rather interesting features:

• The primary circuit is hydraulically unaffected by either the operation of the secondary pump or the hydraulic control aspects (not shown) of the secondary circuit load.

• Dynamic hydraulic effects on the primary are contributed *only* by the positioning of the control valves or balancing valve. In this regard, as the primary system sees it hydraulically, a pumped secondary circuit connection is no different from a simple valve-controlled nonpumped load.

• Like the simple valve-controlled load, the use of a throttling valve has no significance so far as the load circuit is concerned—either accomplishes exactly the same thing.

• The choice of a throttling or a three-way valve is purely a consideration of the source system thermal dynamics and the primary system hydraulic dynamics.

To amplify this last point; from the consideration of the source system, the three-way valve will provide an essentially constant flow

[1]Carlson, G. F., "Hydraulic Systems: Analysis and Evaluation, Part 1," *ASHRAE Journal*, October 1968.

rate to the source with a decreasing temperature differential at reduced load while having a minimal hydraulic effect on the primary system, whereas the throttling valve will provide the source system with a reducing flow rate of essentially constant temperature differential. Given the two alternatives, as stated above, the constant flow variable temperature range alternative imposes a control burden upon the source, particularly if it is a chilled water system, and a significant energy burden upon the primary pumping system.

Three-way valve is choice

In spite of these evident disadvantages, the three-way valve has been the overwhelming choice of systems designers—for both large and small systems—for decades. The reasons: simplicity of understanding the hydraulic impact and assurance of adequate flow rates through the source apparatus. *It is time to re-evaluate this logic*—which brings the discussion to the source systems.

In small systems with a single-source device (chiller, boiler, or heat exchanger), the advantages to be gained by the use of throttling valves may be difficult to justify in terms of the complexity introduced to address the hydraulic effects, and there is no control advantage. However, as system sizes and module numbers increase (two or more), the throttling valve operation option cannot be ignored.

Consider chilled water system

Consider, for example, a chilled water system. If there is one single humidity critical load, the primary supply chilled water temperature must be held at a given design temperature, with control span variation only. However, if the multiple chiller units are piped in parallel and one or more units are cycled off as the load reduces, noncooled return water through the down unit will mix with cooled water from the operating unit, raising the temperature of the supply water. This phenomenon has been found to be one of the major problems in the larger chilled water systems.

Again, the recognition of the hydraulic and thermal phenomena inevitably leads to rather simplistic answers to the complex problems. In this case, the concept of circuiting a pri-

mary loop around the chillers, with no hydraulic impact of the source upon the distribution system and a separate module-assigned source pump, provides a reasonably valid alternative for virtually all multiple-unit systems. This alternative achieves a degree of simplicity that is readily understood, allows numerous alternatives for the designer for methods of interconnecting the load subsystems, depending on other system dynamics, and allows the design of small or large systems that consume minimum process energy while improving performance.

47

Integrated decentralized chilled water systems

Chilled water when applied to air conditioning refrigeration systems is employed as an intermediate heat transfer fluid for the purpose of conveying heat from the space (via a cooling coil) to the refrigeration cycle from which it is normally "pumped" to the higher temperature sink of the outdoor air or an available water source. As technology in chilled water systems developed, it became increasingly evident that in addition to being a thermal conveyor, enabling a physical separation between the load and the source (such that each could be located for the convenient satisfaction of other design parameters), two additional primary advantages emerged:

1) The thermal lag or time constant provided by the inherent storage characteristic of the water provides improvements and simplifications in the control of both the air side apparatus and the refrigeration machinery.

2) When multiple points of cooling or conditioning the air are required, the diversity between these loads can be applied to the refrigeration machinery size and operating modes. This feature results in lower investment in refrigeration and dissipation apparatus and a reduced energy consumption by the refrigeration prime mover.

See growth of central plants

In recent years, the centralization of chilled water refrigeration systems has been seen to

This chapter was reprinted from *ASHRAE Transactions*, Vol. 82, Part I, 1976 and was originally entitled, "Integrated Decentralized Chilled Water Systems." Appearance of this material in *Energy Engineering and Management for Building Systems* does not necessarily suggest or signify endorsement by the American Society of Heating, Refrigerating or Air-Conditioning Engineers, Inc.

grow beyond the lines of a single building, lending to the extended use of central chilled water plants serving shopping centers, campus-type developments for educational institutions, health care facilities, office complexes, and municipal-type plants serving multitudes of commercial customer loads.

In many cases, whether by original planning or for lack thereof, a grouping of buildings such as a college campus, or a single large building, has developed a system of separate chilled water systems serving individual buildings or individual portions of a single building.

The concept discussed herein is a method for integrating these isolated chilled water systems into a single system or "loop" to regain the advantages of the centralized system.

Consider campus as example

Consider, as an example, a hypothetical campus shown in Fig. 47-1. The load quantities represent the full load refrigeration system requirements for each of the ten buildings.

Consider, further, two basic alternative methods for providing for these loads with chilled water systems. First, Serve each the loads from a single chiller located in the respective building. This alternative would require the installation of ten chillers, with a total refrigeration capacity of 2000 tons. This capacity would have to be provided in both the refrigeration prime mover apparatus and heat dissipation apparatus, such as condenser water pumps and cooling towers.

This approach would, under any increments of part load (say, for example, 10 percent load on all buildings) require the operation of all the chillers, each at a greatly reduced load and

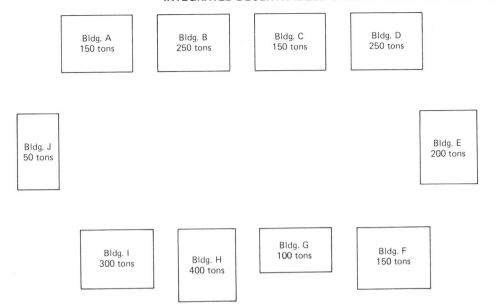

Fig. 47-1. Hypothetical Campus Loads

most additional supplementary equipment at full load. Another disadvantage is that of reliability. Since, in the hypothetical example, each building is provided with one chiller, in the event of a failure of that unit, the building would be without cooling. With current building technology, this degree of reliability in the refrigeration systems is totally untolerable.

The second alternative would be to serve all of the loads from a central chilled water plant. If this were done, the diversity of loads between buildings could be reflected back to the size of the chiller plant. This diversity results from:

1) relationship between time of peak load of buildings resulting from variations in use, functions, architectural design, orientation, etc;

2) shifting of occupants on campus from one space to another.

The studies of numerous campus cooling systems reveal that a building/system diversity of 0.70 applied to the sum of the building loads is fairly representative. Thus, a central plant for this campus could be provided with 2000 × 0.7 = 1400 tons, resulting in a reduction from the individual systems of some 600 tons of refrigeration and heat dissipation

apparatus. The chilled water distribution system required would possibly partially offset this monetary savings. The distribution system would require a basic pipe size capable of conveying the entire 1400 tons as a minimum, and the primary pumping circuit would require the energy input to circulate the 1400 tons of cooling capacity throughout the campus.

The lack of individual building reliability realized with the individual systems, and the numerous hours of operation of multiple machinery, each at greatly reduced loads, are effectively resolved with well-directed design of the central plant. Both parameters are satisfied, in most cases, by proper selection of the machine modules or sizes, selected to match the part load profile of the integrated system. Often, an acceptable degree of reliability can be achieved without the investment burden of excess capacity. Matching the module size to minimum number of hours of operation will generally provide a statistical probability of coincident failure during those few "peak load" hours well within the range normally required in comfort cooling.

Integrated system explained

The integrated decentralized chilled water system concept is an attempt at achieving a

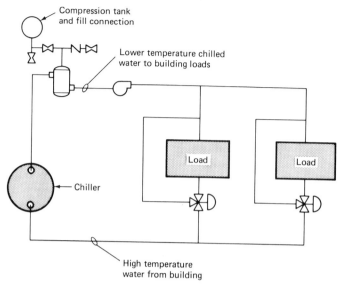

Fig. 47-2. Typical Simplified Building System

majority of the advantages of both the de-centralized and the central plant approaches. The concept is to connect all the building systems into a common pipe or pumped loop. If a typical flow diagram of a building system is as shown in Fig. 47-2, the concept of the integrated decentralized loop is to tie the load into the integrated "system" and provide the capacity of the chiller to the system. For this building, this connection would be made as shown in Fig. 47-3. The development of the

loop for the hypothetical campus would simply extend this methodology throughout the campus to all ten buildings, resulting in the flow diagram shown in Fig. 47-4. Although it may not be immediately evident that full load advantages exist, consider that if the campus diversity is 70 percent, Fig. 47-4 shows an integrated system with an integrated capacity of 1400 tons and a machinery capacity of 2000 tons.

Since all the loads and sources are con-

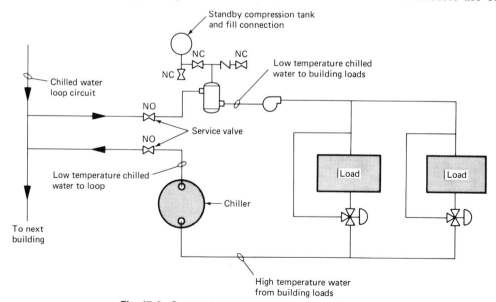

Fig. 47-3. Connection of Building System to Loop main

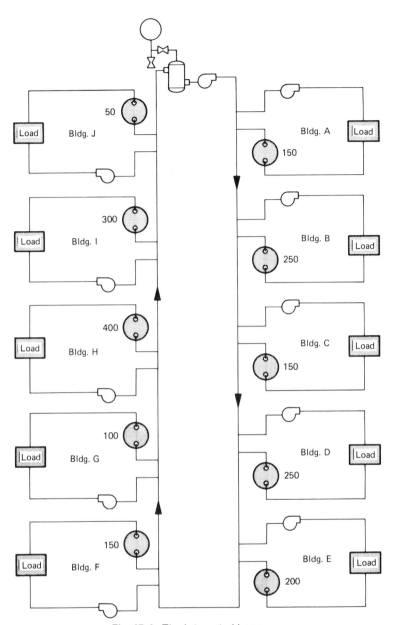

Fig. 47-4. The Integrated Loop

nected in series along the "closed loop" the first step in analyzing the loop dynamics is to perform a temperature gradient analysis. Such a full load loop analysis consists of determining the design coincident load for all of the buildings and the time the coincident design will occur. The results of this analysis in tabular format are shown in Fig. 47-5.

After determining the coincident full load of the connected buildings, the loop design can proceed to selection of the loop size and pumping rate. Note that at this point, assuming the individual systems projected in Fig. 47-1, that we have a singular chilled water system with 2000 tons source capacity and a coincident connected full load of 1400 tons, or a reserve (and "pickup") capacity of 43 percent.

The minimum loop circulating capacity must be matched to the full load circulating

Building	Full load	Coincident diversified load	Chiller capacity	Excess capacity
A	150	50	150	100
B	250	250	250	0
C	150	100	150	50
D	250	100	250	150
E	200	150	200	50
F	150	150	150	0
G	100	50	100	50
H	400	300	400	100
I	300	200	300	100
J	50	50	50	0
Totals	2000	1400	2000	600

Fig. 47-5. Tabular Summary of Loads.

capacity of either the largest load or source. The temperature variation in either increase or decrease along the loop will be:

$$t_l = t_e + \frac{\text{load Btuh}}{(500)(\text{gpm})_{\text{loop}}}$$

or

$$t_l = t_e - \frac{\text{source Btuh}}{(500)(\text{gpm})_{\text{loop}}}.$$

Thus, assuming that all the sources and loads for the campus shown are based upon 12 F, the minimum loop circulating capacity would be dictated by Building H, which would be

$$\text{gpm} = \frac{(400)(12,000)}{(500)(12)} = 800 \text{ gpm}.$$

Under maximum or full campus load, then, a temperature gradient around the loop is developed. The starting temperature is assumed as the nominal design or "target" temperature, and the starting point is irrelevant. A temperature *range* for the entering water temperature in each building must be established, or following the preliminary analysis this range is established. If the arbitrary starting point for the temperature analysis is

set as the entrance to building A, Fig. 47-6 shows a tabular analysis for the temperature decrement around the loop. Note that in Fig. 47-6, 50 F has been established as the maximum loop temperature. Thus, whenever the temperature would tend to exceed 50 F, the capacity of a chiller is provided. A more descriptive method of projecting this information is with a bar chart which is warped to show the loads as diagonal lines downward and to the right and the sources as vertical lines. Such a chart is projected in Fig. 47-7, showing the same data that appears in tabular form in Fig. 47-6. In the figure, the scales on the left ordinate and the abscissa represent loads and chillers in tons of refrigeration, and the right ordinate scale represents loop temperature. Note that if the loop is connected as shown in the flow diagram or connection diagram (Figs. 47-3 and 47-4), the temperature of the loop for any given building will be equal to that on the diagram "leaving" the chiller. The useful loop temperature is always represented at the tail of a diagonal arrow on Fig. 47-7.

For the example shown and developed thus far, the flow rate was selected as the minimum

Bldg.	Entering temp. °F	Loop less chiller °F	Loop with chiller °F	Loop temp °F	Chiller not running tons
A	44.0	45.5	Off	45.5	150
B	45.5	53.0	45.5	45.5	—
C	45.5	48.5	Off	48.5	150
D	48.5	51.5	44.0	44.0	—
E	44.0	48.5	Off	48.5	200
F	48.5	53.0	48.5	48.5	—
G	48.5	50.0	Off	50.0	100
H	50.0	59.0	47.0	47.0	—
I	47.0	53.0	44.0	44.0	—
J	44.0	45.5	44.0	44.0	—

Fig. 47-6. Full Load Temperature Decrement.

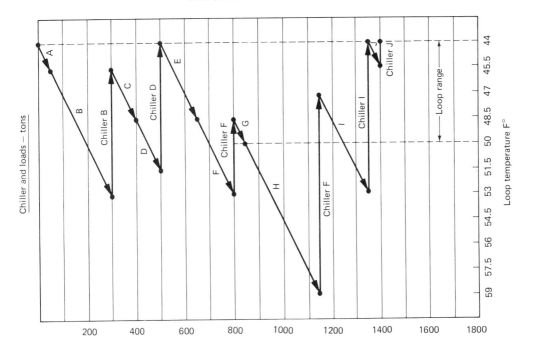

Fig. 47-7. Chiller & Loads Versus Chilled Water Sources.

flow being equal to the flow rate required by the largest load or source connected to the loop. This minimum is rather self-explanatory, since a cursory inspection of the flow diagram will reveal that, at any load, if the loop flow is less than the load or the source, recirculation will occur. The other limit, that of maximum loop flow, is a function of the maximum tolerable building system entering water temperature. The loop range established was 6 F, which was defined by the maximum permissible temperature entering any building, and the nominal design or minimum loop temperature. If a lesser range were desired, (say maximum of 3 F entering any building) the flow rate would simply be determined by the relationship:

$$\frac{gpm_2}{gpm_1} = \frac{\text{loop range 1}}{\text{loop range 2}}.$$

Thus, if the specific design requires a maximum entering water temperature to any building of 47 F, the loop flow rate would be:

$$gpm_2 = gpm_1 \frac{50\text{-}44}{47\text{-}44}$$

$$= 800 \frac{6}{3}$$

$$gpm_2 = 1600 \text{ gpm}.$$

The reduced load operation enables the remaining machines to be dropped off the line as the loop temperature drops. However, the geographic location of the respective loads and sources imposes a limitation on the minimum number of units that can be operated. It is not within the scope of this chapter to illustrate the entire technique of the part load analysis, but two approaches are suggested, both being an iterative analysis of the thermal dynamics versus the various reduced load conditions:

1) Conduct a series of part load calculations including the various occupancy situations and weather conditions, then simply plot the warped bar chart of Fig. 47-7 (or tabular data of Fig. 47-6) for each identified condition.

2) If a computer program is available which calculates the part load, the loop mathematics can be interfaced with these data to provide

the machine operating hours and ton-hours directly.

Caution must be exercised in using the latter approach since most computer programs do not take into account such variables as control ranges, operating techniques, and machines deadlined for overhaul or service.

Explain elements of control

Control of the integrated loop system can be either by manual start-stop or the logic can be automated. However, due to the inherent decentralized nature of the system, remote monitoring of temperatures is almost imperative. As a minimum, the temperature *entering* each building load system should be sensed. As the temperature increases beyond the control range at any point, an upstream source unit is put on the line. Again consideration of relative location of loads and sources along the loop will dictate which units to cycle.

There are some elements of caution which must be exercised in developing a system of this type:

1) Since the entering chilled water temperature will exceed that normally designed for, cooling coils must be selected for the higher entering water conditions. This may necessitate, in some cases, higher flow rates per ton of cooling requirement in the building circuits.

2) Humidity-critical loads such as computer rooms, operating rooms, etc., if incorporated into the loop, must be done with caution. Experience has shown that when used for comfort cooling only, higher than design loop temperatures have been tolerated, resulting in space-humidity levels above design. The operators of the systems have found this mode of operation perfectly acceptable. This mode could not be tolerated, however, if such humidity-critical loads existed.

3) The simplified flow diagrams of Figs. 47-3 and 47-4 are presented simply as an example. However, the feature which is most significant is that, hydraulically, the primary pumping system of the loop must be completely independent from the building or source-pumping systems. To retain this independence, the supply and return connections to each building should be made immediately adjacent to one another with no changes in loop piping size. Furthermore, the loop flow rate should never drop below the extraction rate at any point.

4) In the building connection circuit, the chiller should always be circuited downstream

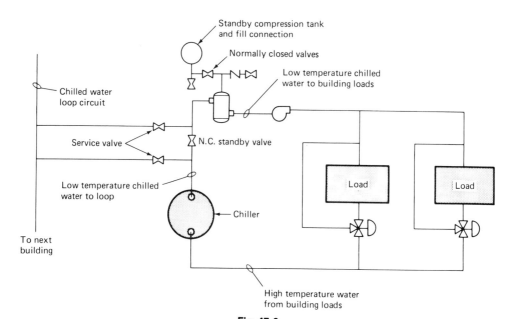

Fig. 47-8.

of the load, i.e., on the end *entering* the loop. The reason is to make available to the chiller the highest possible water temperature.

Aside from these limitations and caution, there are no further differences in the building system piping arrangements from a more conventional chilled water system design. For instance, the building system can be arranged for variable flow—primary, secondary, etc. To achieve energy economics, if sufficiently sophisticated control is provided, the primary loop flow can be reduced with load reduction.

From the standpoint of reliability, this system has the same inherent weak link of any centralized system, in that failure of the primary pumping system or a rupture in the distribution piping would affect the entire campus. Figure 47-8 indicates a method of separating the buildings from the loop under such conditions by the simple addition of a valve, and standby operation compression

tank and fill connection. This feature is not attainable with central plant systems.

Summary

The integrated decentralized loop chilled water system provides many of the advantages of a central plant system with the additional advantages of:

1) standby operation in the event of loop failure;
2) lower investment cost in distribution piping.

It is applicable to:

1) existing campuses or building complexes where the decentralized approach has been taken on a "growth" or piecemeal basis;
2) new campus-type developments where scheduling of building construction and development is such that the central plant cost burden for the initial buildings is not feasible.

48

A case study of an integrated decentralized chilled water system

Solution maximizes utilization of existing chiller capacity and upgrades overall campus air conditioning while reducing machine operating hours and operating and maintenance costs.

The campus of the University of Missouri at Rolla, like many others, reflects the changes in building technology that have occurred over the past half century—a period in which the mechanical and environmental disciplines of that technology were changing at an exponential rate. The problem posed by such a situation is one of periodically reevaluating older buildings from a functional point of view and arriving at decisions as to whether capital should be invested to upgrade these spaces environmentally and functionally or whether they should be abandoned to obsolescence.

The Rolla campus at the time of this undertaking consisted of 30 buildings including the power plant, as shown in Fig. 48-1. Ten are served by modern central chilled water cooling plants; nine of these systems have absorption chillers, while the remaining one has an electric motor-driven centrifugal unit. Total installed chiller capacity is approximately 1817 tons. Steam for the absorbers is supplied from the power plant, which also provides steam for space heating and domestic hot water heating. The low-pressure steam is available as a by-product of power generation. Table 48-1 lists the buildings having central chiller systems.

Cooling in the remaining buildings is provided by unitary equipment, window units, or both. Generally, only partial cooling is provided in these structures. In service are 53 pieces of unitary equipment totaling 424 tons of capacity and 246 window units with a combined capacity of 325 tons. Tables 48-2 and 48-3 show the distribution of this equipment throughout the campus.

A study was undertaken to evaluate the present and future cooling needs of the Rolla campus and determine the feasibility of developing a single chilled water plant. The objectives were to reduce energy consumption, operating costs, and maintenance costs while increasing reliability and maximizing the useful life of the equipment.

Obviously, the existence of approximately 1817 tons of chiller capacity could not be ignored because of the capital investment involved and the long service life that this type of equipment can provide. The challenge was therefore to develop an integrated system making maximum use of the existing central plant type equipment. This led to the *decentralized* central plant concept in which the existing chillers would be tied together by chilled water loops.

An inventory of the existing chiller systems

198

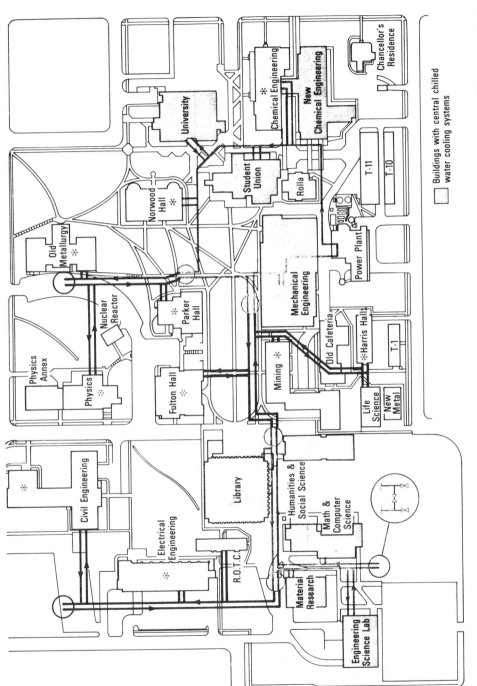

Fig. 48-1. Campus of the University of Missouri at Rolla. Buildings shown by shaded areas have central chilled water refrigeration plants. Shaded piping shows Loops 1 and 2, Loop 1 at left and Loop 2 at right, in proposed construction phases to ultimately provide cooling to most of the campus from the existing chiller systems. Together with the black piping, they make up the Sigma Loop envisioned in the third phase of development. Installed capacity will handle virtually the entire campus because of diversity in building loads. Future piping connections for the initial loops and final loop are shown by the circles on the piping layout. Buildings marked with asterisks are those considered for inclusion in the Sigma Loop system.

Table 48-1 Buildings served with central chilled water systems.

Building	Area, sq ft	Equipment	Installed tons
Physics Annex	14,800	Absorption	71
New Chemical Engineering	78,600	Absorption	360
Library	85,600	Absorption	274
Mechanical Engineering	38,922	Absorption	200
Student Union	17,900	Centrifugal	115
University Center	38,400	Absorption	204
Materials Research	28,600	Absorption	120
Math & Computer Science	35,900	Absorption	123
Humanities & Social Science	30,600	Absorption	155
Engineering Science Lab	42,400	Absorption	195
Totals	411,722		1817

was followed by a thorough analysis of their operating characteristics. The system evaluations provided revealing information that emphasizes the need to take into account energy utilization in environmental systems design.

This chapter deals with the development of the integrated chilled water system. Chapter 23 explores energy economics concepts and shows how the work at the Rolla Campus underscores the need for considering these concepts in all building design programs.

Existing chiller systems

Descriptions of the central cooling systems on campus are given in what follows. Flow diagrams showing major system components were prepared for all of these buildings.

- *Physics Annex*—Its system includes a 71-ton absorption chiller and a crossflow cooling tower with an indoor sump. The air side system includes a 12-ton multizone unit, heating-cooling unit ventilators on a two-pipe distribution system, and fancoils on a two-pipe system. Perimeter finned-tube radiation is provided in some spaces. Space and piping arrangements have been provided for the future addition of another absorption unit with associated pumps and cooling tower.

- *New Chemical Engineering Building*— This building, under construction at the time of the study, is served by a single 380-ton absorption unit. The air side consists of two high-velocity central station type air handling units supplying dual duct variable volume

Table 48-2. Unitary cooling inventory.

Building	Installed tons
Rolla Building	10.0
Physics Auditorium	25.0
Norwood Hall	15.5
Parker Hall	60.0
Fulton Hall	15.0
Harris Hall	37.5
Old Cafeteria	12.5
Old Metallurgy	2.0
Civil Engineering	35.5
Chemical Engineering	18.0
Chemical Engineering Addition	15.0
Electrical Engineering	23.0
Reactor	7.5
Life Science	17.5
T-10	5.0
T-11	20.0
T-C	17.5
Math and Computer Science	87.0
Total tonnage	423.5
Total number of units	53

T denotes temporary buildings

Table 48-3. Window unit inventory.

Building	Units operating	Total tons
Rolla Building	15	21.4
Physics	5	7.8
Norwood Hall	32	54.2
Parker Hall	1	0.8
Fulton Hall	30	37.2
Harris Hall	12	12.2
Old Cafeteria and T-6	2	3.2
Old Metallurgy	6	5.2
Civil Engineering	9	15.5
Chemical Engineering	41	66.0
Mechanical Engineering	2	3.2
Electrical Engineering	21	24.2
Reactor	5	4.1
Life Science	3	3.0
T-10	4	7.8
T-11	4	5.9
T-1	4	2.7
T-2	9	10.7
T-4 (Military Classroom)	4	5.0
T-7 (Military Supply)	3	2.8
Chancellor's Residence	8	7.5
Mining	26	24.4
Totals	246	324.8

terminal devices. The two central station coils are wild-flow devices and represent the only two "loads" on the chilled water system. Computer rooms, which require close control of relative humidity, are provided with self-contained air-cooled units, each of $7\frac{1}{2}$ ton capacity.

• *Library*—The library is served by a single 275-ton absorption unit located in a penthouse equipment room. The air side consists of two high-velocity dual duct air handling units located in a room immediately adjacent to the chiller room. The cooling coil banks in the built-up units, running wild with an uncontrolled cold deck, are the only chilled water users. Heat dissipation is through multiple forced draft cooling towers located on the roof adjacent to the equipment room.

• *Mechanical Engineering Building*—Only the 38,920 sq ft addition to this building, built in 1967, has been provided with central air conditioning. The cooling source is a 200-ton absorption unit. On the air side are 36 ceiling-mounted four-pipe air handling units with chilled water coils supplying terminal reheat coil zones. Supplemental perimeter heating is done with wall hung convectors.

• *Student Union*—This is the building that employs the electric motor-driven centrifugal chiller. It is a 125-ton hermetic packaged unit. The distribution system is a two-pipe hot and chilled water network serving three central station air handling units and 44 terminal fan-coil units.

• *University Center* (new Student Union)—Chilled water to cool the building is provided by a 200-ton absorption unit located in a ground floor mechanical equipment room. Condenser water is cooled by a roof-mounted cooling tower. The chilled water distribution is separate from the hot water distribution (four-pipe) and supplies three multizone air handling units and one small draw-through conditioner. The kitchen makeup air unit is not provided with a cooling coil.

• *Materials Research*—A 120-ton absorption chiller located in a ground floor equipment room supplies chilled water to a two-pipe hot-chilled water distribution system serving room terminal units, which include a combination of unit ventilators and fan-coil units. Outside air is supplied to those units toward the building interior by a nonpressurized central outside air intake duct system. The cooling tower is located on the roof.

• *Math & Computer Science*—Cooling is provided by a 170-ton absorption unit located in a penthouse equipment room. The cooling tower is roof-mounted. The terminal system is supplied by three high-pressure central station air handling units (zoned south-interior-north), which utilize terminal reheat units for

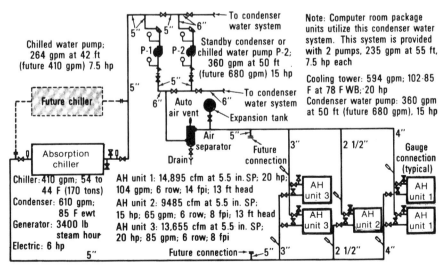

Fig. 48-2. Flow diagram is for central cooling plant in Math & Computer Science Building. It is representative of those prepared for all buildings with central plants.

space control. A computer area is served separately by five water-cooled self-contained units for year-round temperature and humidity control. Computer room units are supplied with condenser water from the same tower serving the absorption unit, but separate pumps are provided. Flow diagram and equipment data for this system are shown in Fig. 48-2.

• *Humanities & Social Science*—Building cooling is provided by a 135-ton absorption unit located in a central grade-level mechanical equipment room. A terminal reheat medium-pressure central station air handling unit is located in the same equipment space. This terminal reheat system provides conditioning for the entire first floor and interior spaces and corridors on the second and third floors. A two-pipe unit ventilator system provides conditioning for the second and third floor classrooms, and the same piping loop serves two-pipe fan-coil units in the offices on these floors. Supplementary heat by convectors and finned-tube radiation is provided where required in lobbies, entries, etc.

• *Engineering Science Laboratories*—The completely air conditioned building is provided with chilled water by a 200-ton absorption unit located in a basement mechanical equipment room. The chilled water is circulated through three central station air handling units that provide space conditioning through high-velocity terminal reheat distribution systems. The condenser cooling water is provided by a roof-mounted cooling tower. Air handling units and all other major equipment items are located on the lower level.

Plan for long range

Discussions with the university staff and inspections of the buildings not currently provided with permanent central year-round air conditioning systems yielded helpful information with regard to long-range planning.

Buildings considered to be of sufficient continuing value to campus operation to warrant upgrading through the addition of cooling apparatus (though remotely in many cases because of building age) were included in the initial integrated campus cooling source analysis. These buildings are so indicated by asterisks in Fig. 48-1.

The Old Metallurgy Building was dropped from the final analysis because of its age and type of structure, which definitely rendered a major investment in central cooling impractical from an economic standpoint. Connections to the loop were shown for this and some other buildings not included in the analysis because of their proximity to the proposed campus distribution routing, however. Added chiller capacity would have to be installed to handle buildings not included in the final analysis.

Load analysis

After the installed air conditioning equipment was surveyed, building cooling loads were calculated to determine how they compared with installed capacity. A detailed computerized load calculating program, available from a time-sharing computer service, was used to generate the load figures. The results are shown in Table 48-4. The buildings that are served by machinery capable of being tied into a loop system have a combined peak load of 1403 tons against an installed capacity of 1817 tons. The arithmetic sum of the peak building loads is nearly 25 percent less than the available tonnage.

It is apparent from these figures that the chillers can handle much more space than they do now since one of the major advantages of supplying a group of buildings from either a central plant or an integrated loop system is the noncoincident nature of the various peak building loads. Load diversity would permit the chiller equipment to serve additional buildings whose individual peak loads added to the present arithmetic load sum would total more than 1817 tons.

The major advantages of the primary loop system are:

• Existing machinery is used as components of the basic system.
• The distribution system (loop) allows for unlimited addition of building loads and

Table 48-4. Calculated cooling loads versus installed capacities.

Building	Area, sq ft	Cooling load, tons	Installed capacity, tons	Load, sq ft per ton
Buildings with central cooling systems				
Physics Annex*	14,800	37	71	400
New Chemical Engineering	78,600	268	360	293
Library	85,600	144	274	594
Mechanical Engineering (1967 addition)	38,922	200	200	194
Student Union	17,900	122	115	147
University Center	38,400	166	204	231
Materials Research	28,600	101	120	283
Math & Computer Science	35,900	87	123	413
Humanities & Social Science	30,600	123	155	249
Engineering Science Lab	42,400	155	195	274
Totals average	411,722	1403	1817	293
Buildings with packaged and window units				
Parker Hall	25,500	54	60.0	473
Mechanical Engineering (1949)	17,200	84	0	205
New Metals Building*	4,050	20	20.0	202
Electrical Engineering	46,200	135	23.0	342
Civil Engineering	61,900	150	10.5	412
Harris Hall	18,250	72	37.5	254
Fulton Hall	29,300	98	15.0	300
Old Chemical Engineering	46,302	141	33.0	328
Mining	25,000	97	0	258
Physics	25,400	98	25.0	260
Norwood Hall	50,900	140	15.5	364
Rolla Building*	12,620	30	0	420
Life Science*	4,704	15	17.5	314
Totals average	367,326	1134	257	330
Total load, all buildings, used in analysis		2435		
Total installed chiller capacity used in analysis			1746	

*Not included in final system analysis. Contribution of Physics Annex chiller considered too small to warrant expense of typing it into proposed loop. Old Metallurgy Building not listed in this table because of prior decision that it would be impractical to install central cooling in it.

chiller sources with campus expansion. This *open-end* design flexibility cannot be achieved with any other concept.

• Investment in piping is minimized because mains sized to carry the summary load are not required.

• Pumping horsepowers are minimized because water flow rates sufficient to carry the summary load are not required.

Loop analysis

In view of the above, the sizing and routing of the piping loop are important factors. The basic concept of the campus loop system is to connect the loads and the refrigeration sources in series in such a way as to allow the loop temperature to rise as it satisfies the loads of several buildings. Then the temperature is again reduced by another chiller to provide for

the cooling needs between that connection and the next cooling source.

To best take advantage of this concept, a thorough analysis of the loads on each building and their characteristics of coincidence with other building loads at each hour of the year is needed. The *design* condition, the time of year that the loop *sees* the maximum load, is then determined.

After the maximum diversified load was determined, a trial and error method was employed to select the loop routing that would result in the minimum decrement of the cooling capability of the loop. The analysis was then performed at the various reduced-load conditions to determine which combination of load/source locations would result in the minimum number of chillers to satisfy the maximum number of loads.

Computer programs for developing energy

consumption and part load performance data were used to determine the various load data and annual hours of refrigeration required for the different loops. It was necessary to modify these programs to obtain the desired information. The programs were run for each building, and the results were then run against one another for the alternate combinations of loop configurations.

Loop split into three phases

The development of the loop was split into three phases for construction budgeting purposes. The first two call for the interconnection of adjacent buildings into two separate loops and placing the chillers into loop operation. The third phase calls for the connection of the two loops and extending the loop to pick up buildings proposed to be served by the existing chillers. The loops in the first two phases are denoted as 1 and 2, and Sigma Loop is the ultimate design (see Fig. 48-1).

The results of the loop performance studies are shown in Figs. 48-3, 48-4, and 48-5, which are actually warped bar charts. The vertical axes represent tons of refrigeration capacity; on the right-hand side are temperature ranges encountered in 8 and 10 in. diameter pipe loops. The horizontal axes represent building cooling loads in tons. The charts represent full load building quantities because system diversity applies to the loop as a whole and not to individual buildings.

Temperature of water entering a building from the primary loop is taken at the tail of the appropriate diagonal arrow. Temperature in the loop after the water has returned from a building that does not have a chiller is taken at the head of the diagonal arrow. Temperature in the loop after the water has returned from a building that does have a chiller is found opposite the head of the vertical arrow. Cooling of such a building is accomplished with water from the primary loop before it enters

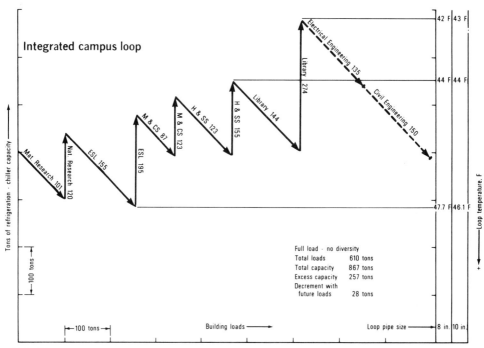

Fig. 48-3. Temperature-capacity gradient diagram for Loop 1 shows performance of Loop 1 (at lower left of Fig. 1). Starting at left chilled water from the loop enters Materials Research Building, absorbs building load of 101 tons (angled line), then gives up 120 tons of heat to the building chiller (vertical line), and re-enters the loop at a lower temperature than when it entered the building. Cycle is repeated throughout the remaining buildings in the loop. Ranges of loop temperatures for 8 and 10 in. piping are shown at right. Dotted lines show effect on loop temperatures if the Electrical and Civil Engineering Buildings were added to the loop and cooled by the excess chiller capacity in the other buildings.

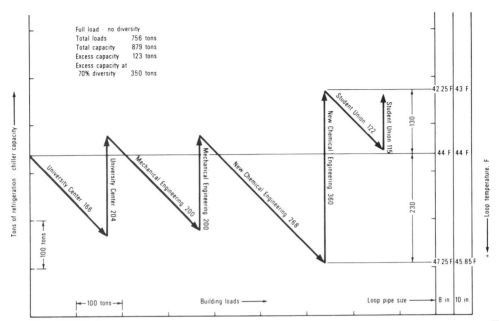

Fig. 48-4. Temperature-capacity gradient diagram for Loop 2 shows performance of Loop 2 (lower right in Fig. 1). Again, angled lines represent building cooling loads, and vertical lines represent installed building capacity.

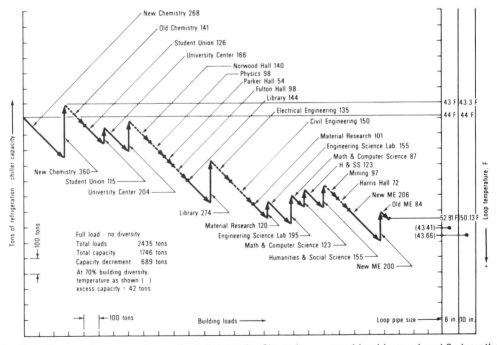

Fig. 48-5. Temperature-capacity gradient diagram for Sigma Loop—combined Loops 1 and 2 plus other connections (Fig. 1). Angled lines are building cooling loads, and vertical lines represent installed chiller capacities in buildings that presently have central plants. Temperature rises in 8 and 10 in. loops are shown at right. Closed loop temperatures with system diversity accounted for are shown in parentheses. Dotted lines show buildings that would be cooled from chillers in other buildings. Not all buildings that have connections to the loop in Fig. 1 are included in the above analysis since the age of some renders economic feasibility of adding central cooling marginal or impractical.

the chiller so that the water entering the chiller is at a sufficiently high temperature to gain maximum capacity from the chiller.

Loops have excess capacity

Examination of Fig. 48-3 reveals that even without allowing for diversity, there is an excess of capacity in Loop 1 since the temperature of water leaving the chiller in the library is lower than the temperature of the water entering the Materials Research Building. Adding the Electrical and Civil Engineering Buildings to the loop utilizes the excess capacity, assuming all individual peak loads are additive. With the 70 percent diversity factor calculated for this loop, however, there is an excess capacity of 240 tons.

Figure 48-4 reveals that at full load, without allowing for diversity, there is an excess capacity of 123 tons in Loop 2. Allowing for diversity, the excess is 350 tons.

Sigma Loop is shown in Fig. 48-5. The 70 percent system diversity changes the capacity decrement of 689 tons when all peak loads are added arithmetically to a surplus chiller capacity of 42 tons. It should be noted that on a diversified basis the temperature leaving the Old Mechanical Engineering Building will be approximately equal to that entering the New Chemical Engineering Building.

Examination of Sigma Loop routing and loading in Fig. 48-1 and Fig. 48-5, respectively, shows that the selected routing provides for the higher loop temperatures at buildings not having central chilled water systems. In these buildings, designers need only select cooling coils and other apparatus on the bases of the available water temperature at the point of connection and the allowable temperature range in the building loop. While these are not difficult restrictions, it is important that they be met if the integrated primary loop concept is to be successful.

It should be stressed that the limiting factor is the maximum temperature at which chilled water can be supplied to a building load from the loop and still provide adequate dehumidification and sensible cooling in the building. Of the two, the latent or humidity control parameter is the more readily affected. The survey of the facilities revealed that those areas on campus requiring close humidity control are provided with independent environmental control systems whose inclusion in the integrated campus system is not contemplated. Thus, the acceptable temperature rise of the loop is directly dependent on the tolerable ranges of relative humidity for normal occupant comfort. Based on the ASHRAE comfort standards, loop interconnection design should be based on a 44 F inlet to the coil with a 16 F range to achieve 75 F DB and 50 percent RH. With entering water temperature and coil dew point swings up to 49 and 60 F, respectively, it was possible to maintain 75 F DB with a maximum rise in RH to 60 percent. Any future critical humidification applications should *not* be connected to the integrated system. All loads that represent comfort cooling *should* be connected.

Temperatures of water entering buildings having chillers are kept as low as possible to minimize changes in existing apparatus. In some instances, additional rows of coils were required. In other cases, there is currently excess coil capacity that can be utilized.

Loop pipe sizing

As stated above, one of the advantages of a loop-type campus system is that the series-connected load and source concept results in minimum pipe sizes for an infinite growth of summary load. Two major considerations in pipe sizing are:

• What is the largest single load or source that will be connected without an accompanying source or load, as the case may be?

• What is the anticipated full load and part load temperature decrement that can be tolerated?

In the study, four basic sizes were considered, 6, 8, 10, and 12 in. The assumed maximum flows of these four sizes are 800, 1700, 3000, and 5000 gpm, respectively. Preliminary analysis led to the rejection of the 6 and 12 in. loops, leaving the 8 and 10 in. loops for further consideration.

The tonnage represented by a load or source

connection to the primary loop is calculated by dividing the product of flow rate (in gpm), 500, and temperature range by 12,000. A maximum temperature range of 16 F in the secondary circuits leads to the following limitations for a *single* load or chilled water source:

- 8 in. primary, 1700 gpm: 1133 tons.
- 10 in. primary, 3000 gpm: 2000 tons.

The flow rates represent the maximum amounts of water that can be economically circulated.

In applying these limitations to the system, there is yet an additional restriction: The chiller source (if the limit is reached) must be applied at a point of maximum temperature; otherwise it cannot be applied at full capacity (16 F). The same limitation applies to a load, calculated as above, except that it must be imposed at a point of minimum loop temperature (43 or 44 F). Thus, in such cases the maximum tonnages above would have to be reduced by a factor depending on the physical location of the connection to the loop.

Figure 48-6 is a diagrammatic illustration of the method of interconnecting a building system with the primary loop. Although differences in the existing chilled water systems will necessarily dictate some custom modifications of this scheme in the various buildings, the proposed connection method can be closely approached in all of the existing systems.

Any future buildings that are planned should include the general piping arrangement proposed, complete with valving and blind openings for the eventual loop connection.

As the final engineering design is developed for each or any incremental phase of system development, some changes to the proposed ideal connection scheme could be dictated by the loop configuration. The basic concept is the relative arrangement of the chiller, load, and loop connection to achieve minimum water temperature to the load and maximum water temperature to the chiller.

Cooling at minimum cost

Summarizing, the previously stated objectives of integrating the presently isolated building cooling systems into a single system are as follows:

- to optimize utilization of machinery;
- to minimize machine operating hours;
- to optimize energy utilization;
- to enhance reliability.

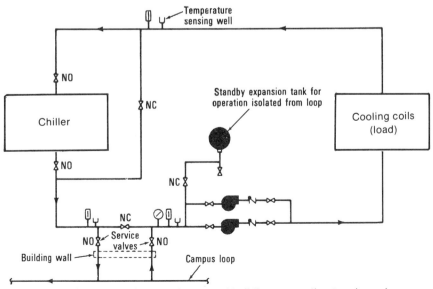

Fig. 48-6. Schematic diagram of proposed building connection to primary loop.

Inherent in the above is accomplishing campus cooling at minimum cost. In this context, cost or economic values are to be applied to virtually all measurable quantities to achieve a common denominator. The quantities involved are:

- useful machinery life;
- machinery replacement cost;
- maintenance costs (planned and unanticipated);
- energy consumption and cost;
- water;
- expendable materials (water chemicals, lubricants, etc.).

To apply a quantitative evaluation to the Rolla campus, the three basic loop developments established were evaluated, each in two different ways: first, with each building system functioning independently of the others—assuming that a central cooling plant would be added to buildings currently not fully air conditioned; and second, with the existing cooling systems integrated into the loop in the three phases of development. This was accomplished by hourly load comparisons incorporated with the reduced or part load characteristics of each chiller system. The following discussions address the energy and operating cost savings achieved.

Energy requirements and savings

Energy consumption was considered in terms of thermal energy expressed in pounds of steam per hour and electrical energy expressed in kilowatt hours. The part- or reduced-load steam consumptions for the absorption units were taken from the manufacturers' cataloged performance curves. Although experience in operating the units indicates that these consumption rates are seldom realized in the installed machines, there was no documentation of this and manufacturers' data were therefore employed.

The comparison was then made for the isolated unit operation and the integrated loop system in the three phases of development, and the results are given in Table 48-5. As can be seen, the annual energy cost saving that can be attributed to total campus interconnection on the Sigma Loop is approximately $10,000.

Machine operation hours and costs

As stated earlier, one of the major advantages of the integrated loop vis-a-vis individual systems is the capability of operating only some of the units under conditions of reduced loading. The value of machine operating hours is less readily identified than that of energy consumption and varies as a function of plant operating and maintenance techniques. The monetary value of machine operating hours is composed of the following values:

- *Capital cost of machinery*—All but one small unit in the proposed system are lithium bromide type absorption units. The basic motivating energy in these units is high-level

Table 48-5. Amounts and costs of electricity and steam for both isolated unit and integrated loop operation in the three phases of development planned.

Condition	Electricity		Steam	
	Quantity, KWHR	Cost, $	Quantity, Mlb per hr	Cost, $
Loop 1 isolated	530,668	9,021	18,821	30,114
Loop 1 interconnected	404,507	6,877	16,291	26,065
Savings	126,161	2,144	2,530	4,049
Loop 2 isolated	524,435	8,915	11,817	18,907
Loop 2 interconnected	339,504	5,772	10,650	17,040
Savings	184,931	3,143	1,167	1,867
Sigma isolated	1,499,343	25,489	45,236	72,378
Sigma Loop	1,055,831	17,949	43,800	70,080
Savings	443,512	7,540	1,436	2,298

thermal energy, which achieves the refrigeration effect through the evaporation of water at extremely low absolute pressures, motivated by a chemical process of brine adsorption. Thus, the "wear and tear" on such a unit is chemical rather than mechanical. It is reasonable to estimate the life expectancy of this type of unit at 20 years, based on 3000 hr of operation per year. Thus, the machine can be estimated to have a useful operating life of 60,000 hr. For simplification, the value of operating hours was developed neglecting value of monies, interest on sinking funds, etc., and the comparisons were made strictly on a cash value basis. Assuming an investment and replacement cost of $250 per ton, average machine size of 181.7 tons, and the 60,000 hr life expectancy, the depreciated value of 1 operating hr of a unit was set at $0.76.

• *Maintenance and operation cost*—Current experience on the campus indicates that the labor and materials costs for the existing units are approximately $28,000 and $12,000 per year, respectively. Calculations indicated that the present operation results in approximately 26,417 machine hr per year; thus, the maintenance and operation value was set at $1.51 per machine hr.

• *Water and water chemical cost*—Although the use of water for condensing purposes is theoretically related to ton-hours rather than operating hours, research indicates that very few data have been compiled relating water consumption to reduced-load demands. It is reasonable to assume, however, that there is a drift loss and controlled bleed loss that exists in a cooling tower system regardless of the load on the system, even assuming a constant temperature-load relationship. This combined loss thus represents a parasitic burden to system operation. In the context of the study, however, the quantitative value of this parasitic loss was not determined, and no value was assigned to this consideration.

Adding the capital cost value component to the maintenance and operating cost value component results in a machine value rate of $2.27 per hr. This rate was applied to the

Table 48-6. Hours and value of machine operation for both isolated machine and integrated loop operation in the three phases of development planned.

Condition	Hours per yr	Value, $
Loop 1 isolated	12,278	27,871
Loop 1 interconnected	5,500	12,485
Savings	6,778	15,386
Loop 2 isolated	11,809	26,806
Loop 2 interconnected	3,700	8,399
Savings	8,109	18,407
Sigma isolated	62,264	141,339
Sigma Loop	13,176	29,909
Savings	49,088	111,430

calculated operating hour schedule shown in Table 48-6 for the three suggested stages of development. The overwhelming significance of the integrated loop approach becomes evident at a glance when the value of the machine hour saving for the ultimate or Sigma Loop is seen to be $111,430 per year. Another interesting correlation is that the machine hours and associated value for the Sigma Loop are *appreciably less* than for the existing nonconnected machinery (by approximately one-half), which indicates that the existing machinery essentially comprising the Sigma Loop apparatus would operate for fewer hours to handle the entire campus than they do now, handling only ten buildings.

Consider sources of inaccuracy

The value of the savings developed above are not without some degree of inaccuracy. The unit value developed is felt to be as accurate as could be generated with the data known, and it can readily be adjusted if desired with further refinement of cost history. The machine operating hours data were developed by calculating the loads for each included building, evaluating the energy and part load profile for each building, and then calculating machine hours. To calculate loop machine hours, all buildings on the loop were summarized as though they responded as a single variable load; then the part load and machine hours were determined. This method of computation resulted in two-sources of inaccuracy:

• Because of the loop capacity decrement, under conditions of extremely low loading, during which the program anticipated only one machine, the temperature rise limitation could be exceeded, necessitating the operation of an additional machine.

• The computerized calculations did not take into account operating tolerances beyond the mathematically ideal condition.

The coupling of these two potential inaccuracies could result in a significant difference between the foregoing predictions and operating experience. If system performance is adequately monitored, however, the deviation should not exceed a 100 percent increase in actual loop machine hours over the number predicted. This correction, applied to the Sigma Loop, would decrease the value of the savings to $81,520—still a very significant amount.

It should also be considered that just as construction costs can be expected to increase in coming years, energy and operating costs will likewise escalate. An analysis of current market trends indicates that both of these costs will increase at a more rapid pace than construction costs during the coming decade because maintenance personnel costs are presently lagging behind construction personnel costs and the current energy shortage will undoubtedly result in significant cost increases in the near future.

Monitoring system

To assist in gaining optimum value from the integrated loop and to retain manageable proportions over the extensive machinery that will ultimately be gathered on campus, installation of a monitoring system was recommended, beginning with the first phase of loop development.

Minimally, direct transmission to a central panel should provide temperature indication at the entrance to each load and each chilled water source; loop pump operation; loop flow indication; and loop pressure indication. Each building could be provided with a failure or "out of tolerance" indicating panel, with a single signal for each building being transmitted to the central console.

Experience has shown that future construction planning for campuses of this nature is at best a fluid situation. Any provisions in a central or integrated energy system that impose a restriction on planning or system revisions can prove economically disastrous. The future growth potential was therefore studied thoroughly, and the results yielded another plus for the integrated loop concept; i.e., since both chilled water sources and loads can be added to the system and their effects controlled completely by the individual building system design engineer, the loop system developed will be completely compatible with any future growth. The only primary loop restriction would be if any such future load or chilled water source, when translated from Btuh to gpm, exceeded the total flow capacity of the loop at the connection point.

49

Preheating outdoor air with transfer fluid systems

The problems associated with the heating of outdoor air under critically controlled temperature requirements need no restating. However, it may help to redefine the premise under which freeze-ups occur in makeup air-heating coils. Hence, this chapter is limited to the discussion of design criteria relating to the heating of makeup ventilation air. The following interrelated parameters must be satisfied:

1) Add heat to relatively large quantities of outdoor air being introduced into an occupied space.

2) Control the temperature of the heated air within close tolerances.

3) Control quality of air within relatively close tolerances.

4) Achieve this process with minimum consumption of energy.

The method discussed herein is the outgrowth of more conventional methods which have been employed over the past few decades, to wit:

1) Heating air by direct firing into the air stream, using staging and modulated firing to achieve temperature control.

2) Heating with high- or low-pressure steam in finned coils, achieving temperature control by steam flow modulation.

3) Heating with high- or low-pressure steam in finned coils, two position steam flow

This chapter was reprinted from *ASHRAE Transactions*, Jan. 1972 and was originally entitled, "Preheating Outdoor Air with Transfer Fluid Systems." Appearance of this material in *Energy Engineering and Management for Building Systems* does not necessarily suggest or signify endorsement by the American Society of Heating, Refrigerating or Air-Conditioning Engineers, Inc.

control, achieving temperature control with face and bypass dampers.

4) Heating with pumped hot water through finned coils using mixing valves for water temperature control with constant flow rates.

5) Heating with pumped hot water through finned coils using fixed water temperature and flow, and face and bypass control.

6) Heating with pumped "nonfreeze"-type fluid (low triple point), using either face and bypass, varying flow, or varying fluid temperature control.

7) Heating with fuel through a combustion chamber-heat exchanger device.

8) Heating with electric resistance coils.

There may well be other alternatives, but those given are the most commonly used methods. Rather than undertake a rigorous review of the advantages and disadvantages, and application and misapplication of each, an analysis with which most readers are familiar, the common criteria always used in selecting one of the above eight alternatives in preference to the others in any design application is summarized. The questions that must be asked are:

1) Will it satisfy the need for maintaining temperature control within limits established?

2) Will it be relatively nuisance and maintenance free?

3) What heating fluid or fuel is available at the point of need?

4) Does the selection justify the investment and operating cost?

5) Does it minimize energy consumption?

The use of these five criteria, with the resulting application to designs of many of the preheater devices mentioned, leads to a value analysis of the alternative systems. In an ensuing effort to find a "better way," this value analyis produced a "hybrid" answer that seems to fulfill virtually all of the design criteria. Such a system would have to:

1) Satisfy the need for maintaining even temperature distribution across the entire section of the intake duct.

2) Minimize energy consumption.

3) Operate with any available fuel, heating fluid, or energy source.

4) Be virtually trouble-free and require minimum maintenance.

5) Provide first cost economics comparable to alternatives.

Needless to say, for air preheaters to be "trouble-free" infers, in addition to normal operational reliability, freedom from "freezing."

The value analysis referred to above was actually stimulated by the growing tendency among systems designers to negate freeze-up

problems by designing complete central hot and chilled water-circulating systems to operate with nonfreeze fluids. The most popular such system is an aqueous ethylene glycol solution. (Some basic characteristics of a 50 percent solution, with a freezing temperature of approximately -3 F are shown in Fig. 49-1.) Applying the appropriate heat transfer, heating capacity, and flow relationships, a significant percentage difference in design and operational energy requirements between this system and a comparable water system is indicated, with a relatively large increase in both heat transfer surface and pumping horsepower as the result. An additional, and perhaps even more significant, problem introduced by this solution to the freeze-up problem is the diligence with which maintenance of the proper level of glycol in the system must be assured. For if this is not done, a false sense of security that the system is "freeze-proof" could lead to disaster in the entire building system.

"Nonfreeze" fluid used

The device selected therefore was a system which actually heats the air through a finned

	45 F		180 F	
	Water	Glycol	Water	Glycol
Specific Heat BTU/lb.	1.003	0.775	1.002	.85
Specific Gravity	1.+	1.07	0.977	1.03
Viscosity (Cp)	1.3	6.0	0.35	0.90
Piping System Δp	1.20	1.70	1.0	1.20

[Flow Rate] lb.m/Time = Heat Requirement/
(Sp. Ht.)(Δt)

Pump H.P. = lb.m/min. × Ft. Hd./33,000

Approximate Increases In Pumping Horsepower
(Not Considering Decreases In Efficiency)

Cooling	Heating
82%	40%

Fig. 49-1. Characteristics of 50% aqueous ethylene glycol solutions versus water.

coil with a "nonfreeze"-type fluid on the inner surface. This coil is close-coupled to a heater section wherein the fluid is heated with whatever fluid or energy source is available. This approach, when applied to a factory-made quality-controlled and tested unit, appeared to satisfy all of the foregoing criteria. A flow diagram showing the basic components and control logic of such a unit is shown in Fig. 49-2 and illustrates the simple concept upon which the proposed solution was based: a nonfreeze-type fluid, fluid heater, and finned heating coil, factory-made and charged to stated specification requirements. The introduction of the intermediate fluid in a hermetically sealed system or assembly satisfies immediately several of the design criteria: (1) isolation of the primary heating fluid from the freezing environment; (2) stabilization and dampening of the control system response through the response time constant created by the intermediate fluid; (3) isolation of the exotic fluid to the closed hermetic system (in preference to a field-fabricated piping system throughout the building).

As is shown on the simplified diagram, the basic control logic concept is a temperature-sensing device in the air stream and directly controls the rate of heat input.

Now let us look at the alternative concepts in the two basic components: (1) the fluid system or cycle; (2) the heater and heat source. The other subsystems or assemblies, including the air heater, control loop or logic, circulator, etc., are essentially dependent on these two, except for an isolated system design which is discussed later.

The single-phase fluid cycle

The first fluid cycle discussed is the single-phase cycle (i.e., no change of phase) which utilizes a liquid throughout the cycle. The basic theory behind operation of the single-phase cycle is identical to that between the so-called gravity circulation hot water heating system or the forced circulation hot water heating system. Continual research is being conducted to determine more desirable fluids. But all fluids which to date have shown favorable characteristics from the standpoint of stability and freeze protection, although showing desirable properties of buoyancy, have been relatively viscous, thus defeating efforts to achieve adequate control response with gravity circulation systems. Thus, the prototype single-phase systems were developed with forced circulation and hermetically sealed pumping devices. Again, the hermetically sealed unit is in keeping with the quality and nuisance control criteria. (Figure 49-3

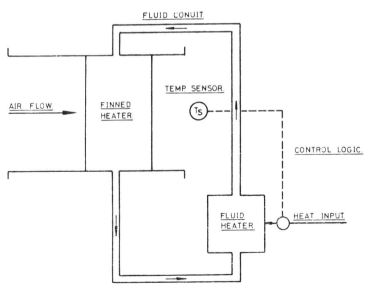

Fig. 49-2. Basic flow diagram.

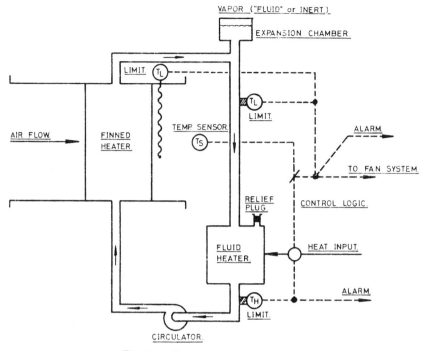

Fig. 49-3. Single-phase pump system.

shows a more fully developed flow diagram of the single-phase system.)

Note that Fig. 49-3, in addition to the circulator, has added an expansion chamber, two low-temperature sensors, one high- or over-temperature sensor, and a relief plug. The logic of the low- or under-temperature sensors is somewhat different from that of the standard freezestat on normal air-handling apparatus. The coil face sensor, upon sensing dangerously low temperatures at the coil face, will override the controlling thermostat in signaling the heat input actuator to accelerate the rate of input (i.e., it will tend to correct the problem by assuming control from the normal controller). When this occurs, alarm contacts are closed to indicate failure on the part of the primary control or heat source. If the correction is not successfully achieved, the "panic button" takes the form of a low-temperature switch on the fluid line leaving the air heater. When this fluid temperature approaches 32 F, the normal sequence of shutting down the fan and closing the dampers occurs. A high- or over-temperature sensor in the fluid line leaving the heater throttles the heat input and

closes an alarm signal to notify of the malfunction.

The two-phase fluid cycle

The second basic fluid system is the two-phase cycle. In this cycle, the principle of the old vapor-type steam system is employed (Fig. 49-4). The concept is to add heat at the heater, or evaporator, evaporating the refrigerant-type fluid which quite logically flows thence to the air-heating coil where it gives up its heat to the air stream. Theoretically, the cycle development could end with this basic concept if the ideal fluid were available; however, a search of fluids to date has not yielded a safe fluid with adequate pressure-temperature characteristics to provide the quality of leaving air temperature control required over the entire anticipated operating range. Thus, as with the single-phase system, a compromise in the basic conceptual simplicity was found to be necessary. Again, the problem was readily solved by the simple addition of a hermetic return pump. Figure 49-5 shows a more fully developed diagram of the two-phase system including its primary control logic.

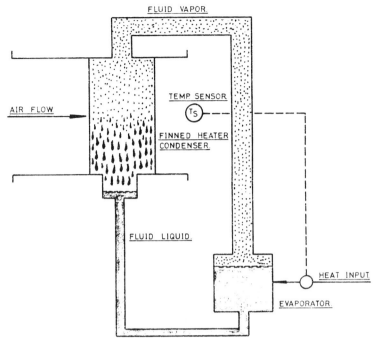

Fig. 49-4. Two-phase simple system.

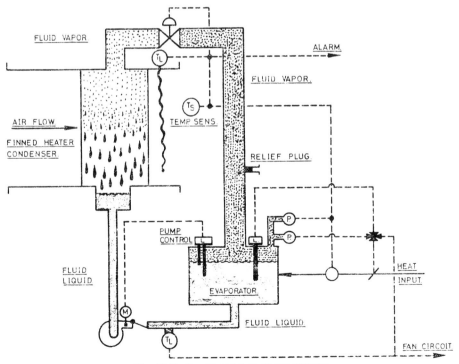

Fig. 49-5. Developed two-phase system.

The control logic of the two-phase system is somewhat more complex than that of the single-phase system. The air temperature control signal positions the fluid valve supplying the air heater and allows the evaporator heater control cycle to function. As in the single-phase system, this circuit is overridden by the low air temperature control on the leaving face of the coil. Evaporator control, initiated by a call for heat from the air sensor, is simply achieved by a variable output or two-position pressure sensor. In the event of overpressure, underlevel, or low fluid temperature, the respective control restricts the heat input; additionally, the low fluid temperature control also stops the fan in the normal "safety" fashion. The pump is simply operated from a probe-type level control. Continued development is being undertaken to bring out the desirable features in the basic concept of the two-phase system. It is strongly felt that some fluid or Azeotrope fluid can be found that will allow use of the high latent heat of vaporization and natural pressure differential flow motivation while dispensing with the complexities of the return pump and fluid control valves.

Primary heating fluid sources

As stated, one of the concepts of the intermediate fluid heater was that it would be applicable to any fuel or primary heating fluid source. Current development has been aimed at five basic heat sources: (1) low- or high-pressure steam (5 to 120 psig); (2) medium- or high-temperature water 210 to 350 F; (3) electricity; (4) natural gas; (5) light oil.

The low- and medium-pressure steam units, whether single phase or two phase, are basically shell and tube heat exchangers with the steam in the shell and the fluid in the tubes. The air temperature sensing controller simply controls the "throttle" of inlet steam valve, and the unit is normally provided with a single float and thermostatic trap or inverted bucket trap with a thermostatic vent port. In keeping with what is believed to be the "coming" field of application in completely closed steam vapor systems, a vacuum breaker is connected from the exchanger to the return line rather than to the atmosphere.

Medium- and high-temperature water units are constructed much like the steam units, utilizing shell and tube heat exchangers and water flow modulating valves controlled by the leaving air temperature sensing controller. With proper sizing of valves to match the unit performance, and built-in stability or stablized time constant, normally one valve is capable of providing flow regulation down to the lowest load requirement. Also, the secondary fluid concept coupled with the dual safety control, provides a more than adequate safeguard against water freeze-up on both the water and steam units.

In the electrically operated system the fluid is circulated through a shell or chamber in which the resistance heating elements are immersed in protective wells. Although this requires appreciably more heating surface than would be necessary were the elements exposed to the fluid, the "well system" maintains the integrity of the hermetic fluid system. The control is quite simple, and an inexpensive method, in that the air temperature controller drives a sequence switch which cycles the heating elements in steps. The introduction of the intermediate fluid dampens the step effect to achieve the end result of variable modulation, thus eliminating the need for expensive solid-state rectifying or clipping of the power wave. More sophisticated high-frequency induction heating from external power probes, with a frequency modulation control, is planned for future consideration.

Probably the most challenging design concept being experienced is the accomplishment of some degree of control and performance for those systems wherein a raw fuel (gas or oil) is the most economical and conveniently available heat energy source. We all think in terms of parameters such as 5 sq ft per boiler HP when considering the problem of heating a fluid via a burner with a fuel. However, some very interesting product research has been conducted over the past decade which resulted in the development of unbelievably compact fuel converter/fluid heaters. Studies currently underway are aimed at incorporating these devices into both the single-phase and the two-phase cycle. The two-phase cycle will utilize a single-phase heater and flash chamber to obtain the vapor.

Consider the heat pipe

Before concluding, consider two other alternate product concepts not yet mentioned. The first one is believed to be the most trouble-free of all the alternative devices discussed, although its application at this stage of development is limited to systems wherein central steam or hot water distribution is available as the energy source for the air heater. On this premise, a unit is proposed which utilizes the refrigerant-motivated heat exchanger known as the heat pipe.

As is widely known, the heat pipe (Fig. 49-6) utilizes a hermetically sealed tube, a wick material, and a refrigerant charge. As one end of the tube is heated, the refrigerant vaporizes, the other end or cold terminal of the tube condenses the refrigerant vapor and creates a low-pressure region to "draw" additional vapor from the vaporizing end. The liquid return system is simply a "wick" or porous capillary material which, by the principle of adhesion, pumps the liquid back to the "evaporator" end. By placing the evaporator end in a "heat source" chamber, and a finned coil condensing end in the air stream, the entire piping and pumping system can be eliminated. Investigatory analysis indicates, however, that the criteria of even-temperature gradient distribution across the intake air duct may not be satisfied, as the gradient along the tube varies considerably. However, it is felt that this problem will be solved by geometric configuration in the not too distant future.

Enter heat-recovery systems

The other product to come out of this development study is one which seems to carry an attractive label these days, that is, heat recovery. Again, from the standpoint of control stability, energy consumption per cfm, and maintenance costs, a highly efficient heat-recovery system has been developed by providing a double-coil system, utilizing the exhaust air coil as the generator or evaporator (as the case may be), and appropriate control logic. However, in comparing energy economics to cost economics, the device is found to be economically unfeasible at this time since no effective efficiency method during the cooling cycle has as yet been achieved. The approach currently explored is that between the conditions of 95/78 and 75/50 percent, 40 percent

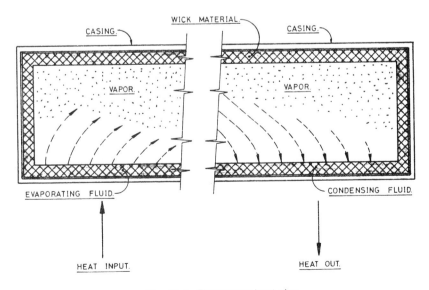

Fig. 49-6. Elementary heat pipe.

of the cooling is sensible; at the same dry bulb temperatures and lesser outdoor wet bulbs (which is normally the case for more operating hours per year), the percentage of dry bulb energy (enthalpy) increases. It is on the basis of this theory that in preliminary studies, tandem heaters (or coolers), utilize the energy source (or sink) from the exhaust air stream, supplemented (in series) with a source (or sink) from a central building system, to provide year-round controlled conditions of the intake ventilation air. Such a system would provide: (1) near optimum energy conservation; (2) an absolutely nonfreeze air preheat system; (3) a precool system which would allow the designer added flexibility in terminal systems control to achieve any desired space-temperature-humidity tolerances.

Heating devices have been previewed in this chapter which will become available in the near future. In the process, a product concept has been provided based on sound engineering principles that will resolve the current problem of coil freeze-up which has plagued both designers and operating personnel for decades. The ensuing savings in time, effort, and actual expense to industry as well as users should prove to be considerable.

50

A state of the art update in steam technology

The inherent characteristics of the two-phase heat transfer fluid phenomena were responsible for the almost universal use of "steam" as an intermediate heating fluid since the origin of central heating systems in buildings until the post World War II era. At that time, advances in single-phase system technology proved to outweigh evident advantages of two-phase systems, and steam application advances essentially ceased. The single-phase concept (water; a water/glycol mixture; or high-temperature low-pressure organic fluids) has predominated system design for 20 years. Many engineers today look upon steam heating systems with the same attitude as they look upon ammonia for refrigeration—only the old-timers possess the outdated skills to design such systems. Furthermore, contemporary designers feel justified by taking the position that single-phase systems are more advantageous in all respects.

Two-phase systems

If, however, one disregards available hardware and current thinking for a moment, the following theoretical characteristics of the two-phase system come to mind:

- Constant temperature heat source.
- Constant temperature heat dissipation.
- Complete isolation of multiple sources and dissipation devices, which enhances reliability.
- Thermally motivated flow.
- High heat content per unit mass flow.

In many applications, these characteristics would make a two-phase system more advantageous than a single-phase one. The next step, then, is to update our thinking on the state of the art of two-phase transfer.

A two-phase system is actually a refrigerant system. As all designers know, with the compressor off, the refrigerant condenses to its liquid phase at the lowest temperature point in the system. The system pressure will be essentially equal to the saturation pressure corresponding to the temperature at that point. The single-phase nature of these vessels results in either subcooling (liquid) or superheating (vapor). If the system design is approached with this basic analogy in mind, the state of the art starts moving forward dramatically. Consider the simple concept of controlling the heat transfer rate from a terminal device, such as a heating coil, heat exchanger, etc. Simplistically, there are two methods of reducing the heat transfer rate significantly: reduce the log mean temperature difference (LMTD), or reduce the heat transfer area.

The first method is used with face and bypass control on a heating coil. As the face dampers are closed, the mass flow of air is decreased, causing the leaving air temperature to rise, which lowers the LMTD and decreases the heat transfer rate. In the limit, heat transfer ceases only when air flow stops completely, and the coil is completely immersed in stagnant air at the steam temperature.

Another common approach to reducing the LMTD is that used extensively in standing radiator systems—the LMTD may be decreased by reducing the steam pressure. The effect of this reduction, however, has significant limitations. Reference to the steam tables and application of elementary heat transfer relationships reveal that if a device is to operate at design capacity of 5 psig, a reduction to 0 psig reduces the heating capability to 90 percent of design. A reduction to 15 in. Hg vacuum (½ in. atm) reduces the capacity to 70 percent of design. When applied to commercial piping system practices, it is difficult to sustain absolute pressures much lower than this. However, if the need were recognized, improved piping systems might be developed.

Heat transfer area reduced

If the heat transfer area is reduced, the analogy of the refrigerant system can be employed again. With low ambient (or air-cooled) head pressure control where a receiver is employed, when the head pressure tends to drop, some of the condenser heat transfer area is "flooded" with liquid. With proper design attention, this principle can be employed with steam heating load devices. As the load decreases, the heat transfer device is simply filled with liquid or condensate. The occurrence of this phenomenon without preplanning in many steam systems has caused numerous problems, ranging from frozen coils to overflowing condensate receivers.

Another method of flooding a steam condenser or heat exchanger is to partially charge it with a noncondensable gas. This approach has been employed as a problem-solving technique to prevent problems caused by liquid flooding. The hardware was quite crude: A simple vacuum breaker on the condenser (or heating device) created an equalized pressure between the condenser and a vented condensate line, allowing the liquid water to be replaced with air. This technique, however, accelerates corrosion in the condensers and condensate systems. If the system designer were to initiate the design with this concept, very likely both problems could be prevented.

Four components of system

The four basic components of a two-phase thermal system are the source (in a steam system, the boiler or steam generator), vapor distribution (steam piping), the load or condenser (heat exchanger, heating coil, absorption refrigeration generator, etc.), and the

return system. Other subsystems or components simply serve as control devices to allow successful operation with varying loads or other dynamic system responses. Return pumps, vacuum inducers, traps, control valves, etc., are not characteristic or necessary basic system components, except as they satisfy the control parameters. As an example, the so-called heat pipe is a functional two-phase thermal system. It utilizes opposite ends of the same pipe as a source and condenser, the center area as a vapor distribution system, and a wicking material as the return (using capillary action forces to motivate the liquid flow from the condenser back to the source).

Development of control and regulation devices or components must originate from a thorough understanding of the uncontrolled input—the load or condenser.

As previously discussed, on the basis of Fourier's equation, extended to steady-state flow in a heat exchanger:

$$q = UA \text{ (LMTD)}.$$

It was mentioned above that two elements of control were the heat exchange area and the LMTD. If one considers how either of these variables can practically be achieved with a control valve, then explores what happens when it closes, two options emerge: The area is decreased by flooding with liquid or noncondensable gases, the LMTD is decreased by reducing the steam pressure, or a combination of the two.

If a system designer selects the option(s) he desires and commences work on the system from that selection, the state of the art update starts moving forward. The solution to this problem (or methodology following the selection) inevitably leads to the return system.

Consider these examples

• *Flooding with liquid*—In past systems, if load reduction was achieved by liquid flooding, the steam trap would open, but when the pressure in the condenser was lower than the pressure in the return line, the fluid (liquid, vapor, or air) in the return line would move into the condenser rather than the liquid moving out. This process would continue until the condenser was flooded with liquid (or air from the return line).

• *Flooding with noncondensable*—The common method of achieving this was to install a vacuum breaker on the condenser such that when the pressure dropped below atmospheric, air would be drawn in through the breaker flooding some of the heat transfer surface. As the load again increased and the valve opened, the air would usually be purged into the return.

• *Reducing the steam pressure*—As mentioned above, the concept of reducing the heat transfer rate by reducing the pressure appears valid in principle; since the condenser component is at saturation conditions, reducing pressure also reduces temperature. However, the limitations in materials and methods seriously restricted the load reduction capabilities, and the result was a combination of pressure reduction and some form of flooding for load reductions below 70 percent of design.

Return piping design critical

When these characteristics are approached analytically, the conclusion which surfaces is that the key to successful design (anticipating the dynamic response phenomenon) is the design of the return piping system and the method of interconnecting the return piping system and the condenser. One finds that virtually nothing has been published in the past 35 years on condensate piping system design, except problem-solving ideas.

Although it may not have seemed obvious to many designers throughout this period, the steam trap is actually a system control device. As such, it has been incorporated into systems with little regard for its actual function, save isolating the steam side from the condensate side. When these traps opened under reduced-load conditions, the result has been far from that anticipated by the designer in many cases (or even by the trap manufacturer). The exact nature of the result depended upon the fluid in the return system into which the trap was connected.

In summary, if the advantages of two-phase thermal systems are to be realized in systems designed in the coming years, in which energy economics, maintenance, and systems reliability are becoming ever more paramount, extensive well-directed research and development in dynamic response, and yes—the lowly steam condensate system—must be addressed.

51

Vapor lock in refrigeration systems

Vapor lock in refrigerant liquid lines is one of the most common causes of malfunction in refrigeration systems used for comfort cooling. Design manuals, such as the *ASHRAE Handbook of Fundamentals* and most relevant textbooks, give little information for the system designer to follow regarding the liquid line design.

As part of an integrated two-phase system, the design procedure for the liquid line cannot be segregated from the physical design of the condenser, receiver, subcooler, and general geometry regarding the relative location of these components.

The vapor lock phenomenon is similar to vapor locking or air binding in a hydronic system. It occurs when the liquid line rises, possibly runs horizontally, and then drops again to the throttling device (expansion valve). In addressing this problem, the first characteristic that must be recognized is that the receiver contains a mixture of liquid and vapor. Thus, the fluid in the receiver is *always* saturated. Neglecting velocity conversion losses ($\rho v^2/2$) and friction losses, if the liquid line leaves the receiver and rises upward, a pressure loss equal to the product of the specific weight (w) and height (h) results. Thus, as the refrigerant rises, evaporation occurs. This phenomenon results in vapor bubbles forming in the higher section of piping.

Pressure differential needed

With the relatively low velocities existent in most liquid lines, the vapor will not continue down the subsequent drop to the expansion valve but will collect in the higher horizontal piping until the line has essentially become totally vapor bound. Subsequently, a significant pressure differential is required to free the line of vapor and reestablish liquid flow to the expansion valve (and evaporator).

As an example, again disregarding velocity conversion and friction losses, if Refrigerant 22 at 110 F (43.3 C) condensing temperature leaves a receiver and rises 13 ft (4 m) vertically, at the top of the riser approximately 13 percent of the available piping volume will initially be occupied by vapor, and to remove this vapor through a subsequent 13-ft vertical drop will require a pressure motivation of approximately 6 psi (41,368.56 N/m^2). But since the vertical column represents a gain of approximately 6 psi (wh), the net motivating force must equal 12 psi (82,737.1 N/m^2). With the net loss of 6 psi, the vapor cannot be recondensed, and thus, the fluid entering the expansion valve will be a liquid/vapor mixture.

Manifestations of this phenomenon in refrigeration systems include:

- *Unstable operation of thermostatic expansion valve*—The valve, sensing excessive superheat resulting from lack of adequate refrigerant, opens full. Following sufficient pressure differential buildup to purge the line, the high-liquid content mixture reaches the valve and the evaporator, subsequently decreasing the superheat and closing the valve. This cyclical action causes surging or instability in the valve operation.
- *Extreme head pressure sensitivity*—Often, with the sight glass (properly placed) immediately ahead of the expansion valve, the only way the glass can be "cleared" of vapor is by overcharging the system until the receiver is

completely flooded, and some of the condenser surface is used as a subcooler. This "correction" results in decreasing the condenser heat transfer surface. In addition to limiting the full load capacity, this makes the system head pressure overly sensitive to changes in ambient temperature for air-cooled units or water temperature for water-cooled units.

• *Significant reductions in system capacity*—A further capacity reduction occurs if the sight glass is located at the receiver outlet rather than immediately ahead of the expansion valve; it will not reveal the liquid vapor mixture existing at the valve location. Thus, although there may be a clear liquid at that point, the fluid entering the expansion valve, being a mixture of liquid and vapor, will significantly reduce the full load capacity of the system.

• *Occasional liquid slugging of compressor*—Generally, the most costly result of liquid line vapor lock is the loss of a compressor due to liquid slugging at reduced-load conditions. With low load on the evaporator, the superheat sensor, as explained previously, positions the valve full open, and following the ensuing vapor purge of the liquid line, the low-quality mixture moves liquid refrigerant through the evaporator and to the compressor before the thermal element can respond.

Useful design guidelines

The most basic approach to follow to prevent vapor lock in liquid line design is to perform a pressure/temperature/quality analysis of the system being designed. Short of this, the following may prove useful design guidelines:

• Remember, if a receiver is employed, the fluid in the receiver is at saturated condition.
• Any useful subcooling must be accomplished after the fluid leaves the receiver.
• Subcooling must be adequate to offset all pressure decreases in all segments of the liquid system (not just, for example, a *net head* difference between receiver liquid level and expansion valve inlet).
• Avoid high points (rise and ensuing drop) in the liquid line whenever possible. Without a high point, flashing can occur, but vapor lock will not result.
• If there are necessary rises in the liquid line, all devices that cause dynamic pressure drop (filter/dryer, solenoid valve, etc.) should be located immediately ahead of the expansion valve.
• Locate a liquid line sight glass immediately ahead of the expansion valve.
• Do not overlook pressure losses resulting from conversion of static pressure in the receiver to flow velocity ($\rho v^2/2$) in the liquid piping.

In summary, this often overlooked phenomenon of the design of the liquid system in a vapor compression refrigeration cycle can either plague designers or be an interesting challenge.

SECTION X

Air systems, adjusting, and balancing

All comfort air conditioning systems have some form of air delivery or transport subsystem, the purpose of which is to move air from the space, condition it, and return it to the space. These air-system types extend over a wide range, from a small propeller fan in a room-type conditioner to fans of hundreds of horsepower connected to literally thousands of feet of ductwork in central station systems. Since there are few, if any, air conditioning systems which do not employ air-circulating systems, it follows that these systems and their components are probably the most common devices employed in building environmental systems. It might thus appear that the technology relating to these subsystems would be thoroughly understood by all systems designers.

Experience in analyzing operating problems in installed systems reveals that this is not the case. Quite the opposite appears to be true. That is, application of fans and proper design of air-ducting techniques appear to be among the least understood subsystems in the overall building environmental systems field. The following scenario might be developed to explain this evident contradiction. First, a simple, limited-extent ducting system appears so simple that an understanding of the physics relating to it seem unnecessary and second, the fan industry was one of the first to apply the techniques of consensus standards to ratings, testing, and configuration.

The combination of these two observations tended to lull the industry into a false sense of security while the systems grew rapidly in their complexity. This growth in complexity and extent was one of the ingredients of the revolution in building technology through the 1950s and 1960s, at a time when general concern for energy consumption was not considered a significant design parameter.

Failure of many systems to perform was blamed on improper system balance. In most cases, the proper balance was never achieved for the simple reason that the systems were unbalanceable. Large air systems are probably the most complex fluid networks of any in current engineering technologies. These large systems contain close to an infinite number of possible parallel or alternative flow paths. Thus, if the systems are to be adjusted to achieve design flow rates to the various branches and outlets, without the generation of excessive noise or the consumption of excessive energy, or both, the methods for achieving this balance must be designed into the system. This element of air-system technology is discussed in the chapter on system balance.

The standardization of the fan industry has had a significant positive effect upon both that industry and the user interface. An unfortunate side effect, however, is that both the industry and the consumers (systems designers) have been too slow to recognize the difference between test ratings and application performance. This subtle but all-important observation is the subject matter discussed in the chapter on the correct use of the fan curve.

The concept of the integration of the fan into the system as opposed to an interface concept, which is an extension to the above problem is explored in depth in the chapter, "Analysis of Fan/System Characteristics and Applications."

Reference to several of the chapters in Sections III and IV reveal that experience in analyzing energy use in existing buildings indicates that the fan energy is often the largest single category of annual energy consumption. An explanation for this is twofold. First, the power requirements tended to be high because of the trend to higher and higher pressures; and second, fans in many systems ran continuously although the buildings were actually used about 20 to 50 percent of the hours of the year.

Consider the energy penalty of high-pressure fan systems. For air at any given condition, the power is directly proportional to the product of the flow rate (cfm) and the pressure rise across the fan (usually expressed in equivalent inches of water). Thus, a system of moderately high pressure, say 6 in. requires three times the fan power as does a low-pressure system at 2 in. There have been many reasons for the trend to higher fan pressures, including pressure required for control of system devices, and pressure resulting from smaller ductwork (for whatever reason). The reasons dictating high fan pressures *must* be challenged if energy-effective designs are to be accomplished—particularly in view of the fact that fan energy is *not* a primary energy requirement but rather is a parasitic burden!

In addition to the direct impact the fan pressure has on the power requirement of the fan drive, it imposes another penalty in a cooling system. The fan energy enters the air system where it is essentially converted to a temperature increase ($\Delta h = C_p \Delta t$). For air at standard conditions and a 74 percent fan efficiency, this temperature increase is approximately 0.5 F for each 1 in. of static pressure rise. Considering that most cooling systems operate through a 20 F air temperature range, each 1 in. of fan pressure relates to 2.5 percent of system capacity. Thus, in a system which operates at 6 in. of pressure rise would utilize 15 percent of the installed cooling capacity, or put another way, it would increase the size of the cooling apparatus by 17.64 percent. As this burden is integrated over all of the hours of operation for the cooling system, *the fan often becomes one of the largest single contributors to the annual cooling requirement!*

Since fan energy is not a primary energy requirement, its input is generally dissipated in performing some function related to making the system work. Many aspects of this performance tend to generate noise or sound which must be attenuated prior to entering the space. This attenuation requires investment in devices increasing the investment cost and since the attenuators impose pressure drops, they in turn increase the energy burden. Thus, if the systems can be designed such that the energy which ultimately generates the noise is not put in initially, the first step toward a quiet system will have been taken.

The brevity of this section should not mislead the reader into assuming the subject matter is either simple in concept or unimportant. The chapter on the relationship between system balance and energy use should direct the reader to the interrelationship between the materials in this section and those in virtually all the other sections of the book.

52
Correct use of the fan curve

Because of the combination of spiraling construction costs, increasing costs of borrowed money, increasing energy costs, and the ever-widening gap between energy supply and demand, the need for more careful attention to the fine tuning of a building HVAC system is rapidly being recognized by both system designers and owners. The efforts at what has come to be called TBA, testing, balancing, and adjusting, are required in new systems at the time of start-up, because the systems currently being designed are less forgiving of maladjustment than those designed in the past, and in existing systems as the first and fundamental step in the retrofit process, whether that process be for the purpose of upgrading (modernizing) an existing system or for instigating an energy management program.

Determine air flow quantities

This chapter addresses only one small part of the system testing problem, but a vitally important part: determining air flow quantities. Much has been published and discussed in open forum on this topic. However, there still appears to be a fundamental misunderstanding on the part of many practitioners. In the vast majority of installations, *the measurement of the pressure rise across the fan, used in conjunction with the published fan curve, cannot be utilized to determine system air quantities.*

The problem is fundamentally that the configuration of the installation differs from that in which the fan was tested and the resulting curve developed. Even with the efforts made in recent years to address the impact upon installation configurations vis-a-vis test configurations, such as system-effect factors, *the fan pressure rise is a totally unreliable metering device.*

Likewise, pressure differential readings across other system restrictions provide, at best, a rough approximation of system flow rate. The reason is similar to that regarding the fan curve: Pressure differentials for such devices as filters and coils are determined by laboratory testing under ideal conditions, wherein the velocity profile across the face of the device is totally uniform. However, in the vast majority of installations, this uniformity is not achieved. Changes in direction, separations, fan inlet or outlet flow patterns are all contributors to the nonuniform velocity profiles.

The only method that provides a reasonable degree of accuracy, then, in determining flow rate in the air delivery system is measuring the velocity or flow rate in the system ductwork at points at which adequate velocity is available for velocity pressure measurements and at which a reasonably flat velocity profile exists. The less consistent the velocity profile, the more points must be measured in the traverse, and the lower the accuracy. It has been said that the majority of air-distribution systems designed to date are actually not balanceable within a reasonable degree of accuracy. If this statement is accepted, the responsibility must be borne by the system designers! In turn, their reason for designing essentially unbalanceable systems has been a lack of understanding of the methods and techniques required for achieving the fine-tuning adjustments.

Correct duct design important

With the foregoing concepts in mind, a fundamental step in the design of any air-distribution system should be to design air metering sections into the air conduit or ducting system. These can vary from flow nozzles to built-in traverse devices to simple sections of straight ductwork of adequate velocity (1500 to 2000 fpm) with gradual inlet and outlet sections. This approach to designing air flow test sections into the system can usually be achieved at no increase in construction cost. It will then provide the ready capability of balancing and adjusting, and reduce man-hours required to accomplish fine tuning.

To summarize, the primary considerations in designing a balanceable air system are:

• Installation of the fan or fan unit in such a way as to achieve the least possible pressure loss resulting from the fan match to delivery system (system effect).

• Routing and sizing sections in each main and branch so that highly accurate flow measurements can be made, preferably by Pitot tube traverse.

• Providing dampers for regulating the flow in each main and branch. The dampers should be mounted in the ductwork. Dependence upon dampers in the terminal devices (grilles or diffusers) is not recommended.

• Attention to the design of the return air or outdoor air intake systems. They are as important as the supply systems.

Once the designer has provided in the system the capability of determining the volumetric flow rate at a point or at points that can be summarized, the published fan curve becomes an immediately useful tool. The initial premise was that the fan could *not* be used for determining the flow rate from a field measurement of the pressure rise. However, once the flow rate is determined, the published curve can be referred to directly to determine the *actual* operating point, in identifying both total pressure and horsepower. This is the subtle but all-important issue in comprehending the significance of curves published for the use of the systems adjustment technician.

53

Analysis of fan/system characteristics and applications

Advanced technology and energy sensitivity in design call for more accurate methods of matching fan to system other than manufacturers' data coupled with safety factors. This discussion, based on actual tests, outlines a way to combine a system curve with a published fan curve for a true picture of system balance.

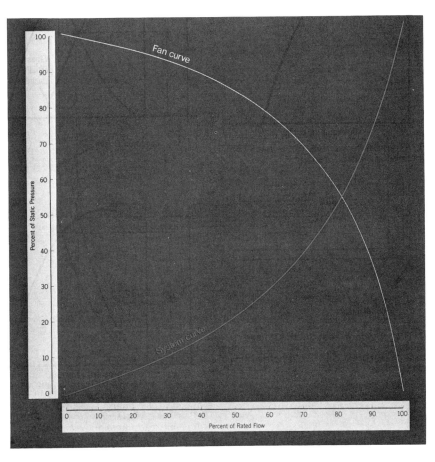

The word "system" and the phrase "systems approach" have been widely used in recent years to describe any number of concepts in engineering and architecture. At the risk of jumping onto a popular bandwagon, in this chapter "system" will be used in two different ways.

First, "system" as defined by the first meaning in *Webster* is "a regularly interacting or interdependent group of items forming a unified whole." It is in this context that systems engineering will be discussed. Second, to paraphrase the definition of system as defined in elementary thermodynamics, the term system "is used to designate any portion of matter that is separated from its surroundings by either real or imaginary boundaries." It is in this context that system will be used to analyze components or component subassemblies that operate within an overall engineered building system.

In the field of building environmental technology, HVAC systems are, in general, designed by a team of systems engineers. The approach is to select a grouping of manufactured products and integrate them together in such a manner that their respective interaction will achieve the design goals set for the overall system. However, over the years, manufacturing technology, taking advantage of the cost efficiencies of mass production versus field erection, has (at the cost of flexibility) resulted in prepackaging of various subsystems to be incorporated into an overall system, in many cases, in lieu of individual machinery components. It is the intent of this chapter to address the subjects of definition of boundaries in order to define the limits of such systems, and the resulting mathematical formulas that will improve the science currently employed by an overall systems designer. Each will be discussed in relationship to handling conditioned air.

Fan curve important tool

Many years ago, manufacturers of air moving devices, blowers, developed an extensive grouping of variables affecting the performance of a blower when applied to an attached distribution system. A product's performance,

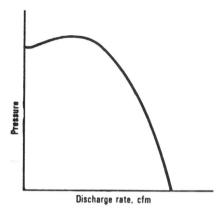

Fig. 53-1. A typical fan curve for a backward inclined or airfoil fan.

as tested in accordance with industry testing standards, is registered as a fan curve. An example of a typical fan curve for a backward inclined or "airfoil" fan is shown in Fig. 53-1. Figure 53-2 illustrates the standard test method for measuring the respective inlet and outlet static and total pressures and volumetric flow rates. It is relevant to point out at this point that a fan is an assembly of components consisting of an inlet cone, wheel, shaft and bearing assembly, and scroll. The purpose of the test arrangement shown in Fig. 53-2 and the resulting curve and rating tables is to provide a standard method of rating the performance of one product compared to another. Members of the systems engineering profession have had only these data to assist in applying a fan to a system since the ratings were established; albeit standard test conditions are seldom, if ever, experienced in actual system installations.

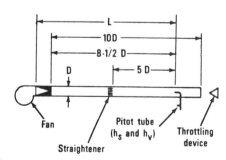

Fig. 53-2. The standard test method used to measure inlet and outlet static and total pressures and volumetric flow rates.

Understand fan laws

An additional consideration of background development are the so-called fan laws. Although much more extensive, all fan laws are based on three basic relationships:

• The discharge rate, cfm, varies directly as the speed.
• The total pressure increase varies directly as the square of the speed.
• The air power varies as the cube of the speed.

These laws are readily indentified by a vector analysis of a fan wheel if the system in which the fan is performing follows the turbulent flow characteristics of:

$$\Delta h = K_s(Q)^2 \qquad (1)$$

where

Δh = head loss,
K_s = system constant,
Q = discharge rate, cfm.

A systems designer, working with the only available information, applies a fan within an air system by using the manufacturer's comparative data in the form of fan curves (or tables), assumes the system curve is based on Eq. (1), plots the parabola against the respective fan curve, and thus identifies an anticipated operating point. This approach, however, has resulted in less than reliable results in a vast number of designs. Heretofore, multiple safety factors incorporated throughout design development have allowed approximation to continue. In recent years, however, with design refinements realized by computerized full load calculations, variable air volume systems wherein full load diversity is applied to fan and distribution system sizing, along with energy sensitivity in design, the profession is faced with the need for more accurate methods for matching a fan component to the distribution and conditioning components of an overall system. One method for assisting a system designer in improving the accuracy of application data accounts for system effects by assigning K-factor constants to various types of fan connection configurations.

Consider total fan system

A fan system is generally assumed to include, as stated previously, an inlet cone, wheel, and scroll as the devices that relate to characteristics; i.e., those devices that if dimensionally changed would result in a different characteristic curve. Thus, a system is defined by the boundaries encompassing these components.

It has also become rather common practice to include within the boundaries of a system variable inlet vanes. As vanes are closed, creating an additional pressure or energy loss at the inlet of a fan, the discharge rate decreases. Figure 53-3 shows a typical method of representing this phenomenon on a fan curve. If two operating points are selected,

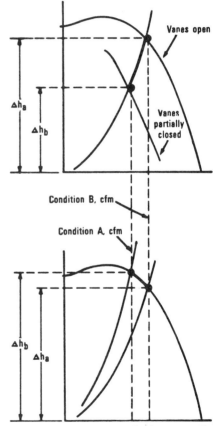

Fig. 53-3. A typical method of representing the decline in discharge rate when vanes are closed, creating an additional pressure or energy loss at the inlet of a fan. Referring to the bottom drawing, the curve moves upward and to the left when a damper is provided on the discharge side and the air flow rate is reduced from Condition A to B.

Conditions A and B, it is readily seen that as the inlet vanes close, both the discharge rate *and* the pressure rise decrease. If, on the other hand, a damper is provided on the discharge side of a fan, and the discharge rate is reduced from Condition A to B, this is generally represented as an increased constant K_s in Eq. (1), swinging the system curve upward to the left. In this case, the discharge rate decreases but the pressure rise increases. What explains the difference in these two relationships?

The basic hypothesis to explain this is simply where the boundaries of a system are defined. In the case of inlet vanes, a damper was included as part of the fan system, and in the other case, it was not. To substantiate this hypothesis, a test stand was constructed, admittedly with considerable liberties as concerns AMCA standard tests. The differences were necessitated by the need for changing the boundaries and obtaining consistent results. Figure 53-4 is a schematic diagram of the test configuration shown in the accompanying photo. The tests revealed that the inlet vane curve could be duplicated by positioning of the discharge damper and vice versa. To remove the inlet vanes from the fan section, the inlet pressure was measured in the center of the fan wheel. In either case, the fan system efficiency (assuming a good selection initially) decreased with reduced flow. However, this reduction in efficiency is less pronounced with the inlet vane system because of the directional nature of the vanes. This improved

Fig. 53-4. A schematic diagram (above) and photograph of the test configuration.

reduced flow efficiency could likely be reproduced or even improved by fixed inlet vanes, which would still allow a designer to extend the boundaries of a fan system to include multiple distribution zones, each with different flow and pressure requirements. The control techniques would be identical to those employed for variable inlet vanes.

The use of inlet vanes for either initial balance of fan to system curve or for operational reduction of flow is not discouraged as a sound practice if it is applied with care. However, an application engineer should be careful in his decision to employ inlet vanes. The idea that reduced flow can be achieved with a more effective reduction in horsepower than possible by the same fan with discharge damper control is not always true. The basic reason is that because of the restrictive nature of inlet vane hardware at a very critical point in the fan system, the tests revealed a significant reduction in fan capacity with the vanes completely open. This restriction dropped the effective fan curve in the test apparatus by a pressure rise reduction of 15 percent with a discharge rate of 0.75 free delivery.

Many devices in air system

A composite air-handling system consists of multiple devices arranged in series with one another. Considering the simple system shown diagrammatically in Fig. 53-5, the devices are: return air inlet, return air duct, filters, cooling coil, fan, supply air duct, supply air grille, and the conditioned space.

In order to clearly understand the behavioral aspects of each device, the components are grouped into various systems by simply defining the respective boundaries. These

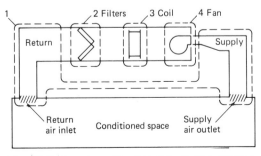

Fig. 53-5. A schematic diagram of the devices included in the boundaries of the test system.

boundaries as illustrated in Fig. 53-5 are: ducting and grilles, filters, coil, and fan.

The reason for grouping in this manner is that the pressure and flow rate characteristics of each device probably differ from those of the others.

The system including the ducting and grilles is found to be described quite accurately by Eq. (1). The source of this equation is the Darcy–Weisbach equation:

$$\Delta h = f(l/D)(V^2/2g) \qquad (2)$$

where

f = friction factor,
l = length,
D = diameter,
V = velocity,
g = gravitational constant.

In this equation, the friction factor is obtained from the Moody diagram and is relatively constant at the higher Reynolds numbers found in air-distribution systems.

This phenomenon of constant friction factor does not necessarily hold true for the flow rate friction loss characteristics of other systems, i.e., the coils and filters. The significance of this deviation from the form of Eq. (1) is becoming increasingly important in systems wherein the major contributors to head loss are these two components. This situation exists with even moderately efficient filtration and deep chilled water cooling coils. It is in this type of system that the classical system curve based on Eq. (1) and even application of the basic fan laws can be misleading during both design and balancing.

Cooling coil pressure drop

The published catalog data for pressure drops through cooling coils reveal that they do not behave in accordance with Eq. (1). The coils produced by one manufacturer were analyzed. The results obtained for wet coils were found to closely follow the relation:

$$\Delta h_c = K_c(Q)^{x_c} \qquad (3)$$

where

Δh_c = pressure drop through cooling coil,
K_c = coil constant.

The exponent, X ranged from 1.66 to 1.81 rather than the factor 2 as found in Eq. (1). For greater accuracy, the coil pressure drop should be expressed as:

$$\Delta h_c = K_{c_1}(Q)^{X_{c1}} + K_{c_2}(Q)^{X_{c2}}. \qquad (4)$$

However, the simpler form of the equation should suffice. Thus, if coil manufacturers would rate cooling coils simply by providing K_c values and X values for each series of coils, a designer could correct the available fan curve to account for coil pressure drop as discussed.

Filter air flow resistance

Filters manufactured of relatively high-resistance tightly woven media have come into common use in large building systems in recent years. These filters and cooling coils often represent the major air flow resistance elements. If a filter and its holding or mounting assembly are considered as the filter *system*, it is found that there are two distinct contributors to air flow resistance: configuration resistance and media resistance. In general, configuration resistance behaves in accordance with Eq. (1), and media resistance follows the laminar relationship of the Hagen Poiseuille law:

$$f = 64/Re \qquad (5)$$

where

f = friction factor,
Re = Reynolds number.

When combined with the Darcy–Weisbach equation, we obtain:

$$\Delta h = K_{f_1}(Q). \qquad (6)$$

If these two relationships were coupled, flow resistance for a filter system would be expressed by the equation:

$$\Delta h = K_{f_1}(Q) + K_{f_2}(Q)^2. \qquad (7)$$

However, with a reasonable degree of accuracy and within the discharge rate limits normally applied, the relationship for filters can be simplified to:

$$\Delta h_f = K_f(Q)^{X_f}. \qquad (8)$$

Again, a literature search of currently available filtration systems with efficiencies ranging from 95 to 35 percent revealed an exponent (X) value of from 1.49 to 1.70 with the higher exponent relating to filters in which configuration loss predominated over media loss.

Filter manufacturers do not catalog filter pressure drop versus flow characteristics for other than clean filters—a most unfortunate shortcoming from the standpoint of applying a filter to a system.

However, if a system engineer accepts a given pressure loss increase as the criterion for replacing filters, this fixed differential can be considered in matching a filter system to a fan system curve. Such a curve could be developed that takes these concepts into account and provides a more accurate picture.

Combining the data

The parabolic curve developed from Eq. (1) that represents the pressure loss characteristics found in duct systems (System 1, Fig. 53-5) is shown in Fig. 53-6. The significance in grouping the data in this manner is as follows:

- A fan system manufacturer could include in catalog data all aspects of a system provided as a product. For example, if a fan is furnished within a cabinet, the resultant losses can be represented as a depression in the fan curve. If the cooling coil is included, this depression in the curve can also be shown, and so forth.

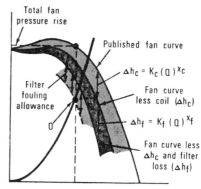

Fig. 53-6. A corrected fan curve that reflects the variable factors of coil pressure drop, filter loss, and fouling allowance.

Thus, for the so-called rooftop units that are currently so widely used, a manufacturer could publish an effective fan system curve to which a systems engineer need only apply the distribution curve to accurately identify the operating point.

• With this approach, the common fan laws could be applied more accurately to assist in system balancing, since the resulting external system follows the second power parabolic relationship stated in Eq. (1).

Consider energy use in design

The sensitivity to energy economics in systems design has created a major thrust to minimize fan power (and resulting energy) that has resulted in lower fan system pressures and an increasing use of variable volume systems. The use of lower system pressures creates situations wherein the major pressure losses are in the cooling coils and the filters, neither of which characterizes in accordance with Eq. (1), leading to error and misunderstanding both in the design and balance phases of system design. The use of variable volume systems results in the need for a clearer understanding of the behavior of the fan within the system—both for reasons of energy control and system performance under varying needs.

Into this latter void has come a groundswell of literature and products portraying the energy consumption and control advantages of controlled variable inlet vanes. In some applications, these devices are well applied; however, a designer should be aware of the limitations of such devices. For example, instead of comparing the reduced horsepower requirements for air flow of an inlet vaned fan at reduced flow to that at full design flow, a designer should first compare the full design flow horsepower of a fan with inlet vanes to the same fan without inlet vanes. In many cases (particularly with smaller fans), the variable inlet vane system may prove to consume more energy per year than the same fan with simple damper control. (However, well-designed fixed inlet vanes may prove *most* beneficial from an annual energy standpoint although these are very rare in the air conditioning products market.)

Standardize product ratings

Unfortunately, the industry rating systems for fan products have been developed for purposes of fair comparison of one product to another—not for the specific purpose of matching a fan to a distribution and conditioning system. The first effort at the latter is the ARI cabinet fan ratings, a system that should be extended to the entire field of manufactured subsystems.

Manufacturers of coils, filters, and other system components should standardize product data on flow versus pressure characteristics in terms of system constants (K) and exponents (X), so that an application or systems engineer could either develop a corrected fan curve to which the system curve could be applied, or an accurate system curve of the form:

$$\Delta h_s = K_c(Q)^{X_c} + K_f(Q)^{X_f} + K_d(Q)^2 \quad (9)$$

where

Δh_s = total system head loss,
K_c = cooling coil constant,
K_f = filter constant,
K_d = distribution system constant.

This equation could then be plotted against the published fan curve.

By combining this actual system curve, the published fan curve, and the mathematical relationship of Eq. (1) with the fan laws, the phenomenon of component matching and system balance would be better understood.

54

The relationships between system balance and energy use

In the study of energy economics relating to energy conversion systems, the first step is to identify the two fundamental components of the analysis, product energy and process energy.

The product energy in building systems is that energy which is required to directly satisfy space environmental needs, in the form needed. Some examples of product energy are the heat introduced into the room at a rate identically equal to the rate at which heat leaves the room to the surroundings and heat extracted from the room at a rate exactly equal to the rate at which it enters the room from the surroundings or relevant internal sources such as the occupants, lighting devices, and appliances.

The process energy is the energy which is consumed by the "system" to satisfy the product need. Examples of process energy are such energy-consuming subsystems as pump energy, fan energy, psychrometric control energy, energy to condition excess ventilation air, and all forms of energy losses.

Product energy and process energy

The total energy required, say annually, by the building is then the sum of the product energy and the process energy.

Experience in analyzing numerous actual buildings and building systems has revealed that the process energy components far exceed the product component in by far the majority of cases—especially if product energy for refrigeration is based on ideal Carnot relationships.

This chapter is directed at the need for careful initial and ongoing adjustments of the various subsystems and components to achieve minimization of process energy. It must be remembered that all process energy is considered energy loss.

Systems technology advances

The field of building systems technology has undergone significant technological advances throughout the past 30 years. The pressures motivating these advances were such that the primary design parameter (and evaluation of success of the design) was the system performance, that is, did it satisfy the product energy requirement.

As a result, very little attention was directed to process energy requirements, and the systems designed were most forgiving of improper adjustments. Some examples of these phenomena are:

1) Oversized primary conversion capacity negated the need for fine tuning the apparatus. Even in a state of poor adjustment, the oversized equipment or subsystem produced the output needed.

2) Low design temperature ranges in fluid transportation or conveying systems enabled most heat transfer elements of the system to perform with flow rates far short of the intended design. An example of this is the flow rate versus capacity curve in the ASHRAE Systems Volume which illustrates that for a 20 F temperature drop coil, 50 percent design flow provides 90 percent design heat transfer.

3) Multiple zoning of high-velocity dual duct air systems, when designed for the sum of the peak flow rates were capable of maintaining space conditions regardless of system imbalance—a control which resulted from the correct proportionate mixture of high- and low-temperature air streams.

Improved design yields results

In the past decade (1965–1975), however, two significant changes occurred. The realization of the limited supply of energy in the form needed for building systems and increased costs of investment monies. These, simultaneous with advances in computer technology (both hardware and software), have resulted in the realization of other design parameters that have been recognized as an equal to that of ultimate performance. This recognition has resulted in the following:

1) Extremely exacting analysis of the product requirement in the form of building design loads, daily time-integrated loads to carefully identify not only peak individual space or room loads but maximum coincident system loads and annual load energy requirements.

2) The selection and design of equipment and subsystems to match the calculated diversified loads.

3) A reduction in the dependency of simultaneous heating and cooling in an effort to minimize psychrometric control process burdens (otherwise known as "runaround" energy).

4) Maximizing temperature ranges in fluid transportation systems for the purpose of both reducing pipe sizes and reducing pumping energy.

5) Variable flow fluid systems (both transport fluid and conditioned air) to reduce annual energy burden.

The systems thus designed are much less forgiving of poor adjustment or balance. As a result, the performance has not been as successful in cases wherein the adjustments were not achieved.

Also, in the earlier systems, in virtually all of the maladjustments which went unnoticed

because of the ability of the system to adjust, such inherent system maladjustments were at the expense of increased process energy consumption.

Two basic requirements

There are two fundamental requirements in the concern of system adjustment and balancing:

1) Initial adjustments. These are the adjustment requirements which must be performed prior to putting a new system into operation. In the case of existing systems which are either being modified or subjected to a newly established energy management program, the initial adjustments must be completed before the management program commences.

2) Ongoing monitoring. These are tests and adjustments which must be made as a part of the continued ownership responsibility in carrying out the preventive or planned maintenance program.

Design guidelines suggested

Providing for the initial adjustments should be the responsibility of the system designer. Unless the system designer recognizes these requirements at each and every phase of both design development and construction monitoring, there is little possibility of successful accomplishment. Suggested guidelines for the designer to follow in addressing this responsibility are:

1) Develop flow diagrams. In each phase of design development, for all thermal fluid subsystems such as air, water, steam, and refrigerants, flow diagrams should be developed prior to the actual layout and modified and updated as the layout proceeds. The flow diagram is the only tool that presents a visual indication of the overall system. Subsequently, all balancing, testing, and adjustment points and devices should be identified on the flow diagrams and they should be included in the construction documents.

2) Develop system schematics for psychrometric and other energy conversion subsystems. These are analogous to the flow dia-

grams for the fluid systems and should be developed for such subsystems as air-handling units, chillers, and converters. The schematic should include such devices as control dampers, control valves, and temperature-control devices. These, like the flow diagrams, should include heat transfer data, flow data, points of measurement for adjustment, and adjustment devices, and should be included in the construction documents.

3) Design a "balanceable" system. With the help of the diagrams the hydraulic and thermal dynamic interrelationships of the system components upon one another at both full load and reduced-load conditions should be identified. It is at this phase of design that "designed-in" problems are minimized. Such potential problems as unanticipated series pumping phenomena, reverse flow paths, fluid overdraw, etc., become evident when the dynamic interrelationships are carefully considered.

4) Assure adequate metering characteristics at measuring points. At all points where velocities are to be used to measure volumetric flow rates, for instance, adequate velocities must be assured and designed into the system. Likewise, for pressure differential measurements, adequate drops must be provided.

5) Construction plans and details should be carefully coordinated with the flow diagrams to assure compatibility and to assure adequate access to measurement and adjusting points.

6) A logistics mechanism for effecting the testing and adjustment program must be achieved. This mechanism is dependent upon the availability of technical expertise in the locale of the building. The technical expertise may be in the form of independent testing and balancing firms, construction contractor, in-house capabilities, or, in some cases this service may have to be performed by the design engineer himself.

Use complete specifications

Following recognition of the available technical expertise, the logistics mechanism must include recognition of the extent of the work to be done when establishing the construction budget, a clear definition of the work and identification of the acceptable technical ex-

pertise in the construction specifications, a mechanism for achieving coordination between the design engineer, the installing contractor, and the testing and adjusting team, and a method for verifying that the work has been successfully accomplished.

Much education is needed

The ongoing monitoring of the system adjustments is, needless to say, the responsibility of the building owner. It must be recognized, however, that the need for this phase of preventative or planned maintenance is a totally new concept to most building owners, if not to many practitioners in the building systems design profession itself. Thus, a good deal of education remains to be done.

This education must proceed as a dissemination of information among the engineering community through our engineering society's technical publications, and seminars along with other available sources. Simultaneously, the informed design professionals should initiate the task of informing their clients, the building owners and managers, initiating at the embryo stages of project description and economic analysis.

The flow diagrams and subsystems diagrams discussed above will prove to be a useful tool in the continued monitoring and adjustments of the system. Design firms with the capability should consider the possibility of providing the services of maintenance management consultation to building owners, and construction contractors with the capabilities should consider the possibility of providing contracted system maintenance which should include the ongoing testing and adjusting.

Identify energy use areas

The preceding discussion was directed toward procedures relevant to new building projects. However, existing building systems which have been constructed without the benefit of this planning should be addressed. These buildings very likely consume more process energy by inherent design than the future buildings will. As a result, there is an ever-increasing demand for energy analysis and resulting steps at reducing the summary energy consumed by their systems. The first step in such an analysis is to develop a mathematical model of the building energy systems which will accurately identify the sources of energy consumption, both product and process. This initial step, the results of which can be verified by historical energy consumption records requires, in most cases, precise testing of the present operating adjustments and maladjustments. Once the operations modes have been verified by field testing, the mathematical modeling currently available through computer programs is acceptably accurate; to a degree that deviations from acceptable norms of accuracy lead to additional field testing which ultimately reveals the area or cause of the deviation.

When this iterative process of calculation and testing achieves compatibility with the operations records, many of the areas of excess energy use will have already been identified quantitatively.

Decide on energy-saving steps

The next step in the retrofit process is to use the mathematical model (the accuracy of which has now been confirmed) to determine the effects of changes to the product requirement (such as alternative fenestration systems), process requirement (such as reductions in total air flow or variable water circulation, and adjustments) on the annual energy consumption. After an acceptable program for modifications in both apparatus and operations techniques has been established, the fundamental requirements of "initial adjustment" and "ongoing monitoring" discussed above should be effected.

Needless to say, any program which does not continually monitor the results to confirm the success of the initial or ongoing adjustment will not have confirmed either the benefits of the effort itself or the degree of success of the responsible agent.

SECTION XI

Maintenance management and reliability

The most fundamental requirement of the effective management of energy systems in buildings is the successful management of maintenance and operation of the machinery and systems.

Management of maintenance requires skills in virtually all the areas normally identified with building business management such as finance, real estate, cost accounting, and personnel; and it additionally requires a thorough knowledge of the technical aspects of the systems. Few building managers possess all of these necessary skills, yet the majority of building systems have operated successfully for years. This evident contradiction should be addressed.

Those systems that have operated successfully with a complete lack of well-directed management have done so at the expense of at least one but usually all of the following:

- reduced level of performance;
- significant increase in energy consumption;
- excessive cost of service and repairs;
- reduced life of the machinery.

The reduced level of performance has generally been used as the ultimate meter of successful building system management because of the ease of identification and record keeping. The measuring device has been the *complaints* or *trouble calls*. A *complaint* is when a tenant or occupant of the building contacts the management and reports a discomfort of the occupants being *too hot, too cold, too dark, too stuffy*, or the like. Since to most managers, complaints are somewhat of an annoyance, the natural and normal reaction has been to minimize the complaints. Complete success in eliminating complaints has been observed as an undesirable goal in some cases. There was one commercial building manager, for example, who had developed a technique for evaluating his own management success on an ongoing basis by relating it to complaints. Too many complaints indicated to him that he was not providing adequate service, and therefore ran the risk of having chronically dissatisfied tenants (which could affect his revenue). Two few complaints indicated that he was very likely spending more money than necessary in providing for the comfort of his tenants. His theory, simply, was that the number of tenant complaints had some point that served to optimize the management economics between revenue and expenses.

Another problem relating to a complete lack of complaints has been observed in building systems management. In the vast majority of instances, the executive level of building management has had a complete lack of knowledge of the technical aspects of mechanical and electrical systems. As a result, they have had to place total reliance upon a maintenance man or staff (consisting of semiskilled technicians). The value of these technicians has historically been measured by their ability to solve problems. If there were no problems (complaints), their value could not be measured. The technicians, in order to be recognized for their true value, needed problems to solve that were visible to both the tenants and the executive level of the management team.

This aspect has another facet which has posed a major obstacle to effective system maintenance. The individual who needs the recognition (for whatever reason) of having solved the problem is psychologically incapable of performing preventive maintenance. There are many such individuals who are totally incapable of preventing problems from occurring, but who thrive on the opportunity to handle the catastrophe once it occurs.

The increase in energy consumption resulting from misdirected management has not only not been a meter of performance, but has not been considered as a controllable variable. Historically, managers of buildings have considered energy costs as an expense beyond their control. A good indication of this attitude is the accounting procedures used in virtually all businesses which contain a building and its attendant expenses. The accounting systems in most cases have placed the energy, if it was purchased from a utility company (gas, electricity, district steam, district chilled water), all in the general category of "utilities" along with water, sewer, etc.; the purchase of other energy commodities such as fuel oil, has often been either included in utilities or in the category of "building supplies" along with soap, deicing salt, etc.

Systems designers in the past have, understandably, been motivated to first, provide adequate size or capacity in the machinery and systems, and second, provide for ultimate control of the temperatures almost irrespective of energy consumption. When the systems were improperly operated or inadequately maintained, the performance was still achieved but at the expense of excessive energy consumption. Examples of this observation are discussed in the chapter, "Lack of Effective Maintenance Causes Excessive Energy Consumption." Since the executive level of management lacked the understanding that something could be done about the excessive energy consumption, this problem was simply not revealed.

The problem of excessive cost of service and repairs is somewhat similar to that of excessive energy use in concept. Most building management teams include a level of executive management (which is skilled in real estate, building finance, and handling problems with building occupants) and a level of technical management which in the majority of cases consists of semiskilled mechanics or technicians whose responsibility is to keep the machinery running. For repairs of major machinery or complex subsystems (chillers, pneumatic controls, prime movers), the technical management people learn to rely upon outside service agencies who unfortunately are motivated to generate accounts receivable by *selling* service. The management mistake is made when the building management team delegates the diagnostic evaluation to the outside service agency. The end result is, more often than not, that service costs are excessive, but management is convinced that these are uncontrollable costs—after all when a chiller is *down*, it must be fixed.

A case history example of this problem is a large high-rise commercial office building which had been spending as much as $120,000 per year for service on a control system. After the management employed a consulting engineering firm to perform the diagnostic services on all machinery and systems, this average annual cost dropped to less than $2,500!

The chapter, "Designing for Reliability," perhaps more than any other in this book, addresses the integration between the contributions of the building manager and the systems designer. On one hand, if the designer does not consider reliability in the design of the systems, the management team will likely have much difficulty in operating the system reliably at a reasonable cost. On the other hand, it is often the building owner-management team who, lacking an understanding of a good engineer's potential impact upon these operating problems, forces the designer to put all these considerations aside and consider *only* performance and investment cost in design decisions. Even more unfortunately, as many buildings are designed, the energy systems designer is isolated from the owner-manager by a third party such as the architect or construction manager who unfortunately may not understand the extreme importance of an intimate interrelationship between the system designer and the owner-operator.

55

Planned versus failure maintenance

A building systems design engineer is now expected to provide considerably more input to a life-cycle cost study of a building venture than he was in the recent past. The current dynamic nature of technology and the state of the art in the building industry have invalidated many of the statistical norms widely used as recently as the last decade. In a life-cycle cost analysis, the three major ingredients controlled primarily by a system designer are: first cost amortization, energy costs, and maintenance/service/operations costs. The latter is discussed here with emphasis on maintenance and service costs.

Maintenance and service costs (M/S) should be grouped into a single entry, simply because well-managed maintenance efforts initiated during the environmental system start-up or debugging phase will reduce service costs during a building's life to a minimum. Conversely, without well-managed maintenance efforts, service costs can be a serious financial burden.

Answer these two questions

Before attempting to quantify M/S needs and subsequent costs, two questions must be answered:

• How long does the owner intend to own and manage the property?
• Are special skills that might be needed for maintenance and service of complex apparatus available in the area, and at what cost?

The answer to the first question dictates whether at time zero a planned maintenance program should be put into effect. If the owner-developer intends to retain the property for a long time, the only intelligent course of action is to instigate such a program. Examples of this type of building are institutional buildings or commercial buildings planned for long-time ownership by major real estate holding firms or sole tenants.

The other alternative, short-term ownership (less than 15 years), might dictate the breakdown maintenance or total dependency on service approach. Whether by plan or not, this latter approach is employed in the *majority* of today's buildings. Examples of projects where this approach is intended are the so-called blue shoe commercial buildings where the owner-developer intends to retain ownership long enough to achieve a financially beneficial crossover point between equity growth and depreciation tax shelter benefits.

The pattern of annual M/S cost versus years after start-up for long-term ownership is shown in Fig. 55-1. If a planned maintenance program is followed, system components will be in as good a condition after a projected 30 or more years as at the beginning. The only deterrent to permanancy is obsolescence.

Figure 55-2 represents a typical M/S cost curve for short-term ownership. Generally, the crossover point occurs between 5 and 12 years.

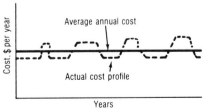

Fig. 55-1. Annual M/S costs of a planned maintenance program.

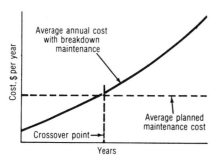

Fig. 55-2. Annual M/S costs of a breakdown maintenance program.

The answer to the first question not only affects a designer's projection of life-cycle cost, but it will also seriously affect the selection of machinery and components. The absence of the answer to this question has in numerous cases resulted in system designers being severely criticized for misdirected decisions. The results of this are that long-term owners are stuck with a property planned otherwise, or rebound owners are not financially prepared for high M/S costs.

Unless there exists a preprogrammed written maintenance program and procedure, it is *most* likely that breakdown maintenance is the operating procedure.

Cite experience examples

The second question is best addressed by a few brief examples from personal experience.

Of the many so-called total energy systems (systems which convert available fossil fuel into all the necessary end-use energy forms for the building system) installed, several have been removed and replaced with conventional forms of available energy. Considering that the concept is valid, once installed, the investment monies spent, and unless extremely ill conceived in energy and load balance, the primary reason these plants have been removed is the unavailability of necessary skills to maintain and service the machines.

An absorption refrigeration system installed in a remote area is another example. A lack of readily available maintenance and service skills led to its removal and replacement by a vapor compression system.

After addressing these two questions, a designer has established the philosophy underlying the M/S aspects of a system design and eliminated certain classes of machines or subsystems. The next step is to assign M/S values to the selection of all system components or subsystems. This is perhaps the most difficult parameter of a system to quantify.

56

Lack of effective maintenance causes excessive energy consumption

It has been said that the best place to hide an object is in the most obvious place because the person seeking it will never think to look there. This saying may also apply to current energy-saving efforts.

In recent years practitioners in the energy conversion disciplines and aware citizens from every walk of life have heard voluminous commentary regarding energy conservation. Government agencies concerned with addressing problems ranging from inflation to unemployment, to gross national product, to national defense; corporate executives concerned with the effects of the rising price of energy on product costs; and homemakers concerned with rising utility bills have all taken steps in their own way to instigate energy conservation programs.

These efforts range from turning the thermostat down to driving no faster than 55 mph, to writing energy conservation standards, to expending millions of dollars on development of energy conservation programs and research aimed at development and utilization of nondepletable sources. An assessment of the potential success of these measures, based upon national depletable consumption statistics, is most disappointing. The least encouraging picture is that of conservation efforts in building systems.

Systems difficult to manage

The reason is that a building system, in addition to being a highly complex energy-consuming entity, is extremely difficult to manage from the energy standpoint. The management responsibility transcends numerous elements of responsibility and technical understanding. As an example, a system designer may accomplish the design of a highly energy-efficient system; but if the installing agency or the operating management do not follow the designer's documents or operating intentions properly in every respect, the system may perform acceptably well but consume significant amounts of energy in excess of the designer's predictions.

It is only natural that for one to look to modifications, retrofitting, investing money to save money, effecting schedule changes, etc. when instigating an energy conservation program in an existing building. These are productive changes easily identified from the standpoint of spending money for a tangible purpose or effecting efforts that can be seen. But numerous experiences in putting energy management programs in effect have revealed that significant conservation can be accomplished by looking in the most obvious place.

Identify system operation

To clear away some of the factors that may be hiding conservation opportunities, the initial step is to identify exactly how the system is operating and then take the necessary actions to achieve operation in the manner conceived by the original designer. During this process, it is generally found that a significant amount of excess process energy is being consumed because controls and the controlled components are either maladjusted or are not functioning in the manner intended. These include such control system features as setpoint, reset adjustments, throttling range (sensitivity), calibration, relay ratios, damper linkages, and actuator operation.

If any system designer were to conduct an in-depth investigation of, say, ten systems he had designed over a span of the past ten years, it would be unusual if he would find one system currently operated and controlled as intended. It it difficult to generalize in establishing causes for this. However, some possibilities may be a lack of understanding (or education) on the part of the operating management, lack of motivation of any responsible agency, and lack of concern or understanding of *systems* on the part of servicing agencies.

Investigations cited in examples

These conclusions are based on numerous investigations of systems serving commercial buildings, institutional buildings, and campus complexes—ranging from systems that have just been started up to ones that have been operating for two decades. Some examples of personal observations are:

• *Damper leakage in multizone units*—A recent investigation of energy consumption on a college campus revealed that a false load on a cooling system of approximately 700 tons was imposed by leakage of dampers in multizone units. The energy to serve this false load was generated by a boiler system consuming about 11 million Btuh of fuel energy.

• *Calibration of controllers*—An outdoor air-sensing controller in a large office building, which indexed the fan system logic from summer to winter mode, was found to be seriously out of calibration. Consequently, the outside and return air economizer damper remained in the 100 percent outdoor air position during all the warmer months while chilled water was being supplied to the coils.

• *Main air leakage in pneumatic systems*—A leak in an air main caused excessive compressor operation that resulted in oil carry-over into the control system. The resulting malfunctioning of control devices led the system operators to respond by essentially aborting the automatic operation and "controlling" a large building system manually at the expense of both excessive energy use and poor performance.

• *Improper adjustment of reset controller*—The improper adjustment caused a wild perimeter heating system to continually overheat during the unoccupied cycle. The operators responded by continually operating the system on the occupied cycle, thus consuming excessive amounts of energy in fan operation and outdoor air heating.

These are just four examples of many, and most have been observed numerous times.

Another germane point is that in all of the above examples (as well as many others), the systems were serviced regularly—some under contract by seemingly responsible service agencies.

57
Designing for reliability

The need to address all relevant design parameters has been discussed in other chapters, in which the general topic either related to design parameters or was more specifically directed to a given subsystem or component.

Reliability is one of the fundamental parameters of the overall integrated building system. And as with any design parameter, the only way in which reliability can be achieved in the integrated system is that it be addressed in the design of all the subsystems and in the selection of all the components that make up the total system. Properly directed sensitivity to the reliability parameter can result not only in vastly improved performance but, in many cases, in reduced investment costs. Improperly directed efforts often result in redundancy and increased costs with little degree of success.

Consider these basics

Consider some of the logic behind these statements.

- The fewer the number of devices a system contains, the greater is the statistical reliability. Thus, the systems designer should always strive to accomplish the required performance with the fewest number of components. Needless to say, there is a strong likelihood that fewer components will cost less money.
- Major machinery should be matched to the anticipated profiles of the system loads. In heating and cooling systems, load analysis generally reveals that the so-called design load is realized in an extremely small portion of the total operating hours. With proper attention to the basic conversion and distribution system design, the components can be selected in such sizes that they can be staged in their

operation to match the load profiles. With this approach, a high degree of reliability can be achieved with little or no investment in redundancy.

Design for demand, base loads

One example may be the selection of water chillers. In a commercial or institutional building, no one designing for reliability would provide a single chiller (although we have all seen cases in which this has been done). One option that has been employed is to provide a redundant or standby chiller also having the capacity to satisfy the design demand load. A far less costly option in most cases would be to install multiple chillers whose summary capacity is equal to the design load requirements. By matching the sizes of the machines to the part-load profile steps, complete standby capacity of one unit could be available all hours of the year except for those few that the last increment of load is required. An additional advantage of the load match increments is that more costly energy-efficient units can be purchased for the base load units that produce the majority of the ton-hours of cooling, and a less costly unit can be applied to the peaking duty. Properly applied, this type of system can achieve the same degree of reliability as could be attained with a system of 100 percent redundancy.

Another example of machinery selection for reliability is in pumping systems. If the fluid system is designed to vary the flow in proportion to load, multiple pumping without redundant or purely standby machinery can be utilized to achieve a high degree of reliability. Fringe benefits are generally a reduction in pumping energy (operating costs) and a reduction in investment cost, the latter resulting from the reduced installed cost of two-way valves in lieu of three-way valves and the obvious cost reduction related to the elimination of standby equipment.

Avoid redundant services

It is all too common for building owners to establish guidelines for reliability for their systems designers based on a reaction to previous problems. In some cases, designers themselves develop such reaction approaches. It is this reaction approach that has been responsible for the concept of redundant apparatus. One of the common examples is the concept of 100 percent standby electrical service, whether this service be by an on-site generating unit or from a second utility service from a different substation. In either case, the cost of the second source is borne by the building investment. This solution, however, although costly, does not protect against faults or failures in the electrical system within the building! With a proper understanding of the problem, an improved degree of reliability can be achieved by the installation of less standby equipment with separate circuiting to

the more critical loads, again with significant reductions in investment cost.

Planned maintenance is vital

If the designer successfully addresses the reliability parameter, the next step in achieving the results rests with the building owner or operator. No system or machinery can be expected to operate reliably unless properly maintained. The term "properly maintained" implies a planned maintenance program. The only other option—breakdown maintenance—by its very wording implies lack of reliability of the broken down device. This, in turn, mandates that the device that will be allowed to break down must be accompanied by a standby unit if the overall system is to be reliable.

Thus, addressing the parameter of reliability inevitably leads to the mature approach of so-called life-cycle costing, integrating the investment cost decisions with those of owning and operating. It is the engineering practitioners who should gain a comprehension of the problem of reliability and give the needed advice and direction to the developers and owners of buildings.

58

Consumer concerns relating to durability, reliability, and serviceability

The concepts of durability, reliability, serviceability, and maintainability are not new to those engineers or businessmen who have been involved in the design and assembly of products since the current technical era began. In the early days of the development of mechanical devices such as the steam engine, automobile, steam heating systems, and air conditioning apparatus and machinery, the primary emphasis was on successful development of a workable machine. Many of the machines from that era are still operational even though technological development has rendered them obsolete. The competitive pressures that existed at that stage of development were overcome by the inventor who developed a better machine. As we pass from the embryo stage in any field of technology into its mature phase, the apparatus or systems that have been developed create a mushrooming market whose need manufacturers and entrepreneurs strive to fill. By this time, the previously available "better ways" inevitably have reached a point of diminishing returns, and competitive pressures are met by placing emphasis on engineering refinement,

This chapter was reprinted from *ASHRAE Transactions*, June 1972 and was originally entitled, "Consumer and Commercial User Requirements As Seen By the Consulting Engineer." Appearance of this material in *Energy Engineering and Management for Building Systems* does not necessarily suggest or signify endorsement by the American Society of Heating, Refrigerating or Air-Conditioning Engineers, Inc.

value analysis, manufacturing techniques, and so forth.

This fact, coupled with the tremendous reservoir of knowledge built up in metallurgy, lubrication technology, thermodynamics, statistical analysis, and accelerated destruction testing, has permitted the manufacturer in today's competitive markets to make product decisions in which he balances "how far can we cut quality" with "and still produce a marketable product." The conscientious manufacturer is trying to compete with the less scrupulous one in a market wherein advertising budgets are often given a higher priority than additional metal gauges or improved quality control—a situation not unique in the air conditioning industry. We all learned in freshman economics, "caveat emptor." "Let the buyer beware" has been true since the first trader tried to outsmart both another trader and a prospective customer. In fields of consumer products which are based on highly sophisticated machinery, the buyer in most instances is not knowledgeable enough to evaluate relative merits between product A and product B. The result is an unfair match in which the consumer is being confronted by a gigantic organization of scientists, marketing specialists, and so forth.

The American Society of Heating, Refrigerating and Air-Conditioning Engineers (ASHRAE) was formed for the purpose of assimilating and disseminating technical information concerning the air conditioning, heating, ventilating, and refrigeration industry—a challenge which the society has coped with to an enviable degree of success throughout a most dynamic period of technological development. Now, the society must, in keeping with its stated purpose, recognize that the area of technical information includes the area of *standards*, with consideration of safety, capacity, and what are discussed here as durability, reliability, serviceability, and maintainability. How can these latter parameters be separated from the former? What good results may be achieved to standardize methods for determining and publishing capacity ratings if a device will only function, say, at the indicated capacity for a short initial period of time? In other words, what happens if the device lacks durability? I maintain that it is absolutely imperative for ASHRAE to pursue the four parameters of durability, reliability, serviceability, and maintainability.

Product quality must be considered

I am associated with a consulting engineering firm; I have an engineering education, and I was virtually raised in the industry. I can boast some 20 years experience in different involvements in the industry. This has enabled me, just like other systems designers or so-called consulting engineers to make a reasonably good evaluation of a product, assembly, subassembly, or component. However, this evaluation is, for the most part, qualitative rather than quantitative, and in many cases is partly based on the reputation of the manufacturer, since information on the components and subassemblies of the finished product is not available. And let me be quick to say that it is frustrating, though not unusual, for consultants to misjudge the durability, reliability, serviceability, and maintainability aspects of a product. At times it can lead to a very sticky or embarrassing situation. For, aside from being a systems designer, the consulting engineer is also an "evaluator of machinery." That is, one of his primary responsibilities is to evaluate and judge products on their relative merits of cost and the mentioned four characteristics which, if considered as an integrated parameter, are normally referred to as "quality."

The fact that ASHRAE formed a task group to study the possibility of a quantitative standard or standards of measurement for these parameters first came to attention through the program for the Society in a recent meeting. The topic greatly appealed to me because for the past few years I have been teaching a graduate course in HVAC systems design. I include in each lecture as a subtopic "comparative evaluation." The usual parameters of cost, energy economics at full- and reduced-load conditions, life expectancy, reliability, and availability of skilled technicians for maintenance and service inevitably generate such student response as the question, "How does an inexperienced engineer make such an evaluation?" The science of functionability, capacity, design feature analysis, and operating cost analysis is rather well defined (except for the part- or reduced-load energy consumption). However, the only reply that I can give an inquisitive student is that aside from the comparison of metal gauges and other superficial parameters he can only base his evaluation on experience. In essence that puts him and his potential client on notice that his ability to evaluate is a direct function of experience; thus he is without a yardstick, and there is virtually no way he can measure these four parameters. It is about time we gave him that yardstick.

Suggested basis for standards

Without going into a lengthy discussion on the advantages of these standards for the ultimate consumer, the scrupulous manufacturer, the contractor (bidding in a competitive market), or the commercial user, the following comments are offered as conceptual seeds on how such standards could be established by ASHRAE and utilized by the consultant who is presumably a knowledgeable buyer. Not all evaluation and measurement techniques can be mentioned here. What is offered is just a preview of the possible areas. For example, whenever an attempt is made to categorize, the decision of delineation must be made, i.e., do we align vertically or horizontally? A vertical delineation would be to take such parameter as serviceability and assign it a listing of various devices. The same can be done for reliability, and so forth.

In a horizontal delineation, each product or device is considered, then the rating methods for that product are evaluated. For the sake of argument, let us consider a horizontal delineation for a single product, also one of the more simplistic product components. It is needless to point out that the more compound the product, the more complex the analysis. Take an air-cooled condenser as an example. Assume further that an air-cooled condenser consists of a coil, a housing, and a fan. For this device, it is desirable to have a means of measuring qualities of serviceability, durability, reliability, and maintainability.

Serviceability as proposed should be a means of measuring or evaluating the performance of the "rated" capacity under actual conditions to be expected in normal use. Thus, the serviceability could be expressed as a safety factor on heat transfer calculations (what in water-cooled apparatus is called fouling factor). Thus, we could simply rate this unit as having a serviceability of 1.1, 1.25, or whatever, so that the consumer can measure the safety factor or compare service expectancy to competitive products.

Durability is more difficult to "rate" since gauges of materials, such as fins and tubes, become secondary when manufacturing ingenuity is applied. However, the measurement of durability must inherently provide a method of evaluating the product in its ability to withstand the abuse of wear and tear as well as normally anticipated abuse which would result in damage and shortened usefulness. Thus, for any given family of products, the standards agency would establish a listing of the yardsticks for the manufacturer to achieve. Such yardsticks might be: (1) protection of aluminum fins from abuse resulting in flattening; (2) protection of resilient mountings from drying up or hardening (due to ultraviolet decay); (3) ability of heat transfer surface to resist fouling erosion or corrosion; (4) protection of electrical components from moisture damage. A set of measurement techniques would then be established, and each of the yardsticks rated independently on a scale from

1 to 10. The durability rating would be the average of these measurements. This would enable the agency to add relevant measuring parameters, or remove those that become obsolete, without changing the meaning of the rating.

Reliability, as I see it, is a double-jointed parameter and easiest to determine after some experience, or knowledge of service life, since the matter is simply one of statistical compilation. All major manufacturers have, or should have, this data on their products. However, this can be looked on as statistical evaluation before the fact, for when a product is developed or proposed, it is most difficult to predict reliability in such a completely revolutionary or new basic device. However, in a new arrangement of components or subassemblies, a simple statistical search of the components and subassemblies should provide an accurate prediction of the unit statistics.

Back to our example of an air-cooled condenser. Considering the subassemblies and components, the statistical reliability of the heat transfer surface and coil would be the experience of the given manufacturer concerning joint leaks or ruptures on similar coils. Similarly, other devices such as motors, bearings, supports, relays or contactors, either would have a statistical history or could be subjected to accelerated statistical failure tests. In the statistical rating method, the weakest link would necessarily determine the strength of the entire chain, and the rating would be determined by the failure history of the product, brought to a base of 10. Products too new to achieve a rating would be so rated until the necessary information is available.

Maintainability is perhaps the simplest preventive parameter no less important than the others. Every manufacturer should be required to provide in all engineering application catalogs and in installation instructions contained in product shipping kits the following information: (1) dimensional clearances needed around the device or product for effective service and maintenance access; (2) component parts listing by exploded view isometrics and part numbers; (3) accurate and easy-to-read maintenance instructions.

For example, on our air-cooled condenser, if clearance is required adjacent to and above the unit to remove the protective screen and clean the condenser, such dimensional clearance should be indicated; likewise for the removal of the fan assembly above the unit.

Clear maintenance instructions

The maintenance instructions to be passed on to the owner should not be multiple choice, such as "If yours has one of these, do this," but rather should apply to that user's specific product. Such instructions might simply be that monthly, during the season of use, the following should be done: (1) inspect the heat transfer surface for dust or other obstructions and clean, if necessary; (2) inspect electrical relay or starter contacts for pitting or corrosion and replace or polish as necessary; (3) inspect resilient mountings and lubricate with silicone lubricant (Ajax No. 123), if necessary; (4) check lubricant level in oil cups on motor bearings and lubricate (with No. 10 S.A.E. lubricant), if need is indicated; (5) inspect belts for tightness and wear, and adjust or replace, if necessary. What I am proposing is not to have a subjective rating on maintainability but rather to certify the item as an approved rating on maintainability, if the maintenance requirements are accurately stated by the manufacturer. The buyer can then apply his own evaluation.

These concepts are not entirely new. The heat exchanger manufacturers for years have had their serviceability evaluation in the form of a fouling factor (although admittedly this has been abused by some unscrupulous manufacturers in our industry). Durability evaluation has only been available in the industry as a matter of judgment on the part of the consuming public. More often than not, in the interest of "the safe side" philosophy, it resulted in unjust evaluation of some manufacturers' products. Reliability statistics have been available only to the manufacturers where they have been used as an effective tool in evaluating available components. However, the buying public has been kept ignorant of this valuable data. Maintainability information is made available upon special request, but usually only in the multiple choice configuration.

I am not naive enough to believe that a product evaluation or rating system could come about in the near or even not so near future. However, such a system appears to be an absolute necessity if the industry is to survive as a bastion of free enterprise. But if the seed is planted now, a full-grown tree will eventually reach its full maturity. I propose that ASHRAE lose no more time in planting this seed, for the sake of stabilizing the industry and protecting consumers as well as manufacturers of integrity.

SECTION XII

Evaluating the effectiveness of energy utilization

The concept of evaluating the effectiveness of energy utilization in conversion apparatus is not new. It was this challenge which commanded much of the attention of the nineteenth century thermodynamicists. The results of these activities are the well-known statements of the Carnot principle that no heat engine cycle can be more "efficient" than a reversible cycle and that no refrigeration cycle can have a coefficient of performance greater than a reversible cycle.

In these definitions, the terms efficiency and coefficient of performance are exactly defined, and using these exact definitions and a bit of elementry thermodynamic relationships, the maximum attainable efficiency and coefficient of performance, respectively, are determined and are found to be a function of the temperature limits of the cycle. In theory, then, these relationships have had two very fundamental uses in the development of thermodynamics and the machines based upon this development.

1) Since the maximum attainable effectiveness is based upon the temperature limits (high and low), a continual effort has been made over the years to develop machinery to operate at the most favorable attainable temperatures. This effort is one of the germane reasons for the economy of scale or size in electric power plants.

2) With given temperature ranges, the major efforts in cycle design have been to develop cycles that come as close as possible to a reversible cycle. This activity has produced the approximations to the Carnot cycle that account for the most commonly used power and refrigeration cycles today, the Rankine cycle and the vapor compression cycle.

In addition to pointing the way to "most favored" cycles, the concept of Carnot efficiency can be used for the basis of comparing the success of any given design. Such a comparison is made in some detail in the chapter, "Thermal Effectiveness of a Vapor Compression Cycle," where the results show that the commercial system analyzed used approximately five times as much energy as the Carnot coefficient of performance would require. Thus, the real cycle could be considered 20 percent effective compared to the ideal.

As is discussed in the chapters of this section, it is essential that in the continued development of power and refrigeration cycles, as well as in pure thermal cycles, the concept of evaluation of effectiveness of energy use be expanded well beyond the simple thermal efficiency and refrigeration coefficient of performance as defined in the days of Carnot. That is not to say that the Carnot concepts should be discarded; quite the contrary, they should be expanded. As used in their classical manner, thermal efficiency and refrigeration coefficient of performance have been *power*, not energy evaluation functions.

As real machines are developed, there emerges the need to supplement the pure cycle with auxiliary devices of various kinds. These auxiliary devices include such system components as control actuators, pumps, fans, etc., all of which consume energy. Thus, at design loading, the true value of thermal efficiency or refrigeration coefficient of performance should consider the burden of the auxiliary devices.

Even more significant is the time-integrated concept which transforms the evaluation functions from power to energy functions. For example, a steam Rankine cycle power plant may have a design capacity thermal efficiency (work out/heat in) of 38 percent but when integrated over a long cyclical time span such as a year, the efficiency may well drop to near 20 percent. A consciousness of the time-integrated nature of energy-use effectiveness will significantly change the approach to system designs. This concept is discussed in the chapter, "Thermodynamic versus System Efficiency."

A bit more subtle is the problem of evaluating the effectiveness of energy conversion when several different useful forms are produced from the same cycle. As, for example, when the heat rejected from a heat engine can be put to beneficial use in such a way that it replaces heat which would otherwise be provided by the consumption of energy resources (otherwise high-level energy).

By the "zeroth law" concept, considering a "scale" of energy, if the reject heat is at a useful temperature it must be above the temperature of an ultimate sink such as the atmosphere. Since the electric energy will also ultimately flow into the ambient sink, the electricity and useful thermal energy can be considered additive. The addition process could be thought of as adding apples and oranges with the sum being so many pieces of fruit.

The problem becomes infinitely more complex when a third product is produced by the cycle in the form of "cooling." So-called cooling energy falls on the "zeroth law" scale below ambient temperatures, such that it will not degrade as did the useful heat and electricity. This concept is the essence of the discussions on energy effectiveness factor.

Attention is also called to some of the chapters of Sections III and VII. In Section III, the chapter, "Proposed Format for Organizing the Study of Building Energy Economics" is included as an effort to provide a method of measuring the effectiveness of energy use in all the elements of a building system, and extending the segregated analysis to the integrated system.

Evaluations of energy-use effectiveness often lead to not only the most desirable or beneficial system from an energy consumption standpoint but also to those which are the most cost beneficial. This is not, however, always the case. Thus, to perform one analysis without the other can and has led to serious errors in judgment on the part of designers.

59

Thermodynamic versus system efficiency

Thermal efficiency and *coefficient of performance* (as it relates to refrigeration systems) are exactly defined terms in classical thermodynamics and as such are well understood by mechanical engineering practitioners. Perhaps it is in the misapplication of these concepts that they have worked to the detriment of effective energy utilization.

A scan of virtually any elementary thermodynamics text reveals two interesting facts:

1) The application of *thermal efficiency* or *coefficient of performance* in examples or problems addresses power rather than energy in the equations (the terms are Btuh, ft-lb per min, horsepower, etc.). Although the definitions and development of the equations are based on energy, a time span must be included in applying the equations—thus, the substitution of power for energy.

2) The "work in" considered is generally always primary motivation power. That is, for a refrigerating machine, it is the work into the compressor. Other work requirements for the cycles are generally ignored. Similarly, for the power cycle, the "work out" is generally considered as that off the turbine (or engine) shaft, neglecting any "parasitic" work input requirements necessary to make the cycle function.

To apply these concepts properly, one should keep the limitations in the use of these terms in perspective. The designer of a refrigerating machine works at trying to set the end temperatures as favorably as possible to achieve the highest possible Carnot coefficient of performance. The next challenge is to construct a machine that will (at full load)

approach the Carnot value as closely as possible. For this reason (and others), it would not be prudent to tamper with our age-old definitions.

Apply new definitions?

Perhaps we should define some new evaluation function and in the process of this definition recognize the limitations of the present definitions. It is suggested that the new terms be *system thermal efficiency* and *system coefficient of performance*, with the *original* terms then being preceded by *thermodynamic* (i.e., thermodynamic thermal efficiency and thermodynamic coefficient of performance).

The system thermal efficiency would then be defined to include all manner and form of energy inputs to the system over a long time span such as an operating season or a year. The basic difference is that the long time span would make the evaluation more truly one of energy rather than power, and the inclusion of the secondary or auxiliary inputs would provide a true measure of the *energy* efficiency. Relating this concept to thermal systems provides a striking example of the value of carefully applying the "system efficiency" concept. In case history studies of active solar heating (and heating-cooling) systems, it has been found that when the system efficiency concept is applied, some of the solar systems are considerably less efficient in the consumption of fossil fuel energy than some so-called conventional systems. The reason is that the shaft energy requirements of the auxiliary devices required to collect, store, utilize, and control the relatively low-grade energy, integrated over, say, 8760 hr per year, are vastly more extensive than those required for some more conventional systems.

Similarly, for refrigerating systems, the definition of the *system coefficient of performance* would include *all* energy inputs integrated over a long time span. Examples of some of these inputs, in addition to primary (compressor) energy, are the energy requirements to drive condenser water pumps, condenser fans, cooling tower fans, and chilled water pumps. These are all part of the energy requirement to remove the heat from the space. A study of numerous existing installations has revealed that in many large chilled water plants, the energy required annually by the auxiliary devices surpassed the energy consumed by the compressor drive(s).

Failure to make a proper distinction between the classical thermodynamic terms and the proposed system terms has led many practitioners in both the private and public sectors to pursue unsuccessful concepts relating to more effective energy utilization. In pursuing these courses, the proponents stopped at the thermodynamic function and concluded that the problem had been solved. It was only after the expenditure of excessive funds that the concept was proven to be a failure, and even then many did not recognize the reason!

Time integration is important

The time-integration feature of this concept is probably more meaningful in the building systems sciences than in any other field of energy systems. The reason, of course, is the combination of the continually varying climatic conditions and uses of the space. Few persons would leave their automobile for the day and leave the engine running; however, it is *most common* that when we leave our building, we leave the mechanical building systems running. In the applications of active solar energy systems, there is little coincidence between the availability of the solar energy and the need. Also, there are many hours (the vast majority in most climates) during which the need for energy (heating or cooling) is but a small fraction of the design value, yet the parasitic machinery is operated at *its* full-rated energy consumption.

Another area where the time integration is of the essence in a basic understanding of the potential cycle effectiveness is in the concept of cogeneration. The thermodynamic efficiency tends to reveal a potential combined cycle efficiency of, say, 80 percent. This assumes, however, a need for shaft (electrical) energy coincident with a need for thermal energy—a coincidence that seldom exists. Thus, before the "concept" is validated, the coincidence must be understood; and then, considering the energy requirements for storage, utilization, etc., the system efficiency can be determined.

In summary, in addressing the energy effectiveness of building systems, the consumption rates must be integrated over a long (repeatable) time span such as a year or a system season, and *no* energy-consuming devices relating to the systems can be ignored. When these factors are considered in thermal, shaft energy, or refrigerating systems, we propose, the evaluation functions of thermal efficiency and coefficient of performance should be clearly identified as different from those in classical thermodynamics by use of the prefix word *system* with the former and *thermodynamic* with the latter.

60

Thermal effectiveness of a vapor compression cycle

The second law of thermodynamics states essentially that heat cannot be made to flow from a region of low temperature to one of higher temperature without the input of energy from an external source. This sets the ground rules: A building enclosed in higher temperature surroundings cannot be cooled without consuming energy. The corollary to the Carnot principle takes us a step closer to quantifying the dictates of the second law by stating that no refrigeration machine operating between a constant temperature source (T_0) and a constant temperature rejection sink (T) can have a coefficient of performance (CP) high than a reversible cycle. A simple thermal analysis of the Carnot cycle (a reversible cycle) further reveals that the CP of the Carnot cycle is:

$$CP = \frac{T_0}{T - T_0}.$$

This is the Carnot or ideal CP. All too often, the tendency is to blame Carnot for our second law energy conversion losses or nonproductive burdens. Consider a simple analysis of how well we are doing compared to the ideal energy requirement. An example will start with 1 ton (12,000 Btuh) of cooling and proceed through an analysis of the useful (product) and nonuseful (process) energy flows through a simple refrigeration system from the cooling coil (input) to the ambient surroundings. For the analysis, conditions are set at 75 F DB and 50 percent RH indoors and a 95 F ambient sink. The Carnot energy required to accomplish heat pumping from 75 to 90 F is 449 Btu; the Carnot energy required to dehumidify, reducing the low-temperature sink to 55 F dew point, is 483 Btu.

The sum of these two ideal energy inputs is the theoretical or Carnot energy required to produce 1 ton of cooling; i.e., 0.273 kW. Another interesting observation is that theoretically, dehumidification alone requires 7.5 percent more energy than the sensible cooling.

Continuing through the analysis, the energy losses or burdens are:

- *Cooling coil heat transfer burden*—The increase in Carnot energy required to provide for a 10 F temperature differential between conditioned air dew point and 45 F entering water temperature is 256 Btu.
- *Chiller heat transfer burden*—The increase in Carnot energy required to provide for a 5 F temperature differential between the leaving (45 F) chilled water and the evaporating refrigerant at 40 F is 132 Btu.
- *Condenser heat transfer burden*—Current product catalog literature was scanned to find a "typical" condensing temperature for a 95 F ambient, air-cooled chiller-condenser combination. This was found to be 121 F condensing temperature. Thus, the heat transfer burden, imposed is represented by the temperature differential 26 F (121–95 F). Incorporating this in the Carnot CP produces a condenser heat transfer burden of 624 Btu.
- *Fluid and thermodynamic cycle burden*—The ideal vapor compression cycle closely approximates the Carnot cycle; the deviations being the superheat resulting from isentropic compression diverging from the vapor dome and the constant enthalpy expansion in lieu of an isentropic expansion. When these deviations are considered with a given refrigerant, the burden of that particular fluid can be determined. For this analysis, the

most common air conditioning fluids, R12 and R22, were considered. The R12 burden was found to be 596 Btu and R22 was 637 Btu.

• *Mechanical and electrical burdens*—At this point in the analysis, Carnot necessarily yielded to available hardware. Again, a product catalog scan and selection of one of the lower kW per ton chiller-condenser combinations revealed that the burden imposed by mechanical and electrical cycle motivation was 1486 Btu.

• *Condenser motivation energy*—In addition to the Carnot heat transfer burden, energy is required to move the cooling fluid through the condenser heat exchanger. Product catalogs were again consulted for a "typical" but conservative value, and the condenser fan resulted in a burden of 471 Btu.

• *Fluid distribution burden*—Since a water chiller was selected for the analysis, the fluid-motivation energy should be considered. Based on a 50-ft pumping head, it is calculated to be 122 Btu.

Table 60-1 shows Btu per ton-hr values for each energy use and burden, and the percentage of input represented by each. Summing all the energy components reveals an input of 4660 Btu per ton-hr or 1.36 kW per ton. If the Carnot CP energy requirement is considered as the ideal, and the thermal effectiveness is defined as the ideal energy requirement divided by the actual, the system is found to be 20 percent effective. Figure 60-1 illustrates the same data in a classical first law thermal flow diagram.

The purpose of this brief excursion into the

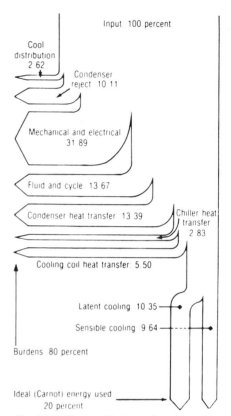

Fig. 60-1. Thermal balance flow diagram.

thermal effectiveness of a refrigeration cycle is to illustrate the theoretical potential for energy consumption reduction in a major subsystem of the building environmental system. If, for instance, the energy effectiveness in heat transfer technology and application could be doubled, this would decrease the consumed energy in refrigeration systems well in excess of 10 percent. Other areas of concentration indicated are in mechanical and electrical technology and thermal fluids.

This analysis has considered only the design capacity energy use and burdens. As the building system operates at reduced loads, the seasonal energy effectiveness is seen to generally reduce radically.

Thus, the question might be asked, would it be more cost and energy effective if we devoted some of the vast sums of money now being spent on such projects as solar energy conversion to more mundane efforts, such as heat transfer, fluids, mechanical concepts, and electrical convertors?

Table 60-1.   Thermal requirements per ton-hour.

	Btu	%
Carnot sensible cooling	449	9.64
Carnot latent cooling	483	10.34
Cooling coil heat transfer	256	5.50
Chiller heat transfer	132	2.83
Condenser heat transfer	624	13.38
Fluid and thermodynamics	637	13.67
Mechanical and electrical	1486	31.88
Condenser heat rejection	471	10.11
Cooling fluid distribution	122	2.65
Total energy input	4660	100.00

61

Energy-effectiveness factor

A new evaluation function permitting comparison of alternate integrated conversion systems.

Throughout the engineering sciences, the concept of evaluation functions is employed as the prevailing technique for comparing alternatives. And in the design of machinery, components, or systems, the iterative process of organized comparisons of alternatives is the fundamental method of ultimately achieving the best result from the alternatives considered. Some examples of evaluation functions include temperature, specific heat, specific weight, enthalpy, modulus of elasticity, thermal efficiency, coefficient of performance, and many others.

This chapter poses the need for, and presents the concept of, a new evaluation function called *energy-effectiveness factor.*

The need for the development of a concept such as energy-effectiveness factor was created by a combination of the awareness of limited energy resources to provide for the needs of an increasingly energy-dependent society and the increasing attention being given to the concept of integrated energy conversion plants (cogeneration systems) or energy communities. For the purpose of this discussion, an energy community is defined as a system within defined boundaries having energy needs in one or more of the following forms:

- Shaft power.
- Light.
- Heat loss to surroundings, requiring an equal input to maintain steady-state conditions.
- Heat gain from surroundings, shaft

power, light, or biological cycles, requiring an equal removal rate to maintain steady-state conditions.

The energy effectiveness factor, Ee, is defined as a dimensionless ratio that enables the effectiveness of the conversion of energy from the depletable resource potential form to the final use form to be expressed.

In efforts preceding the origin of the integrated conversion plant concept, evaluation functions directed at comparing or illustrating the effectiveness of the conversion included:

- *Thermal efficiency*—the net work delivered by a heat engine to some external system divided by the heat supplied to the engine from a high-temperature source, in consistent units (dimensionless decimal or percent, generally not time integrated; i.e, a power ratio at design output).
- *Heat rate*—usually applied to electric power generating plants and defined as the annual or seasonal fuel input in thermal units (Btu) divided by the plant electrical product in kilowatt-hours (Btu/kW-hr).
- *Fuel rate*—usually applied to internal combustion prime movers and defined as the fuel input in gravimetric or volumetric units divided by the shaft or delivered energy output such as (lb fuel/hp hr).
- *Coefficient of performance*—a power function relating to refrigeration machines and defined as the ratio of the rate of heat into the refrigeration system from the low-temperature sink to the motivating power input, both

expressed in consistent units (dimensionless ratio). A dimensional form of COP is called the energy-efficiency ratio, EER, with the numerator expressed in units of Btu per hour and the denominator in watts (Btuh/W).

• *Conversion efficiency*—a function relating to energy conversion in plants wherein the conversion is from a potential contained in input fuel or resource to a high-level thermal output, defined as the useful thermal form output divided by the potential combustion heat input (HHV) in consistent thermal units (a dimensionless ratio usually expressed as a percent).

Develop energy effective use

Integrated plants have been developed through efforts to make more effective use of resource energy potential. They include plants with the following combinations of products:

- Electricity and heating.
- Electricity, heating, and cooling.
- Electricity and cooling.
- Heating and cooling.

In evaluating the effectiveness of these energy service facilities to provide for the needs of any given energy community, a method of comparison of alternative schemes is mandated. If an evaluation function method is used for the comparison, the same method can be employed to compare the performance of the operating entity to the performance anticipated during the early decision-making stages of design.

How to determine Ee

To develop the concept of energy-effectiveness factor, *Ee*, consider two plants that provide only one form of product energy. The first plant, illustrated in Fig. 61-1 by the first law balance diagram, is an electrical generating plant. Definition of the boundaries is extremely important, and for the purposes of this discussion the boundaries are defined as the plant itself and its distribution network to the point of product delivery. What occurs within the boundaries (the "system") in the form of energy conversion characteristics of the subsystems and components contributes totally to the value of *Ee* but is not relevant in the final analysis of the numerical calculation.

Since the energy flow from the "system," *E*, and the product delivered, *e*, are both from the plant to the community,

$$\text{product} = \text{energy flow}$$

$$e = E.$$

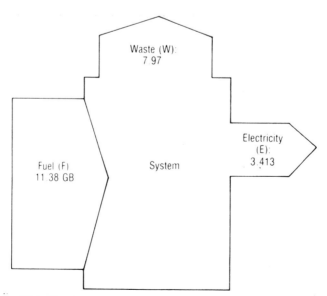

Fig. 61-1. First law balance diagram for electric conversion "system."

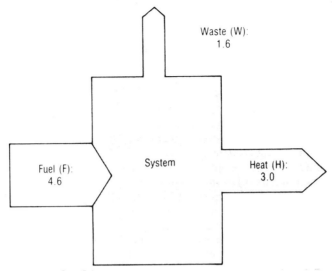

Waste (W):
1.6

Fuel (F):
4.6

System

Heat (H):
3.0

Fig. 61-2. First law balance diagram for thermal conversion "system."

In the example, the annual product delivered is 1 million kW-hr, which expressed in equivalent thermal units is 3.413 GB.*

The input fuel energy, F, is expressed as the high heat value of the depletable energy resources (fossil or nuclear fuel) consumed by the system annually. Since this energy flow is also in the same direction as the product flow, f,

$$f = F.$$

And for the example shown, $F = 11.38$ GB.

The Ee is calculated by dividing the output *product* by the input *product*, or

$$Ee = e/f$$
$$= 3.413/11.38$$
$$= 0.30.$$

The analysis to determine the input energy required to generate the useful output consisted of highly sophisticated calculations involving plant machinery characteristics, annual load profiles, part load subsystem performance profiles, plant burdens, combustion losses, distribution system losses, etc.; but once the input required to produce the annual output was determined, the only data required to calculate Ee were the input quantity and the

*1 GB = 1 × 10⁹ Btu.

delivered output quantity. Also, the only instrumentation required to determine the Ee of an operating plant is metering of the annual input and product.

The second example is the plant illustrated by the first law balance diagram of Fig. 61-2, which is a thermal plant supplying steam to the energy community. The thermodynamic difference between this and the first example is that we now have a first law conversion plant instead of a second law conversion plant. But as in the first example, both the energy flow, H, and the product flow, h, are from the plant to the community; thus, again,

$$\text{product} = \text{energy flow}$$
$$h = H.$$

In the example shown, the product delivered, h, is 3.0 GB, and the input product, f, is 4.6 GB. The energy effectiveness is then:

$$Ee = h/f$$
$$= 3/4.6$$
$$= 0.65.$$

In calculating the energy-effectiveness factor, all inputs to the plant must be expressed in thermal value of fuel. Thus, any electric energy input to the thermal plant of Fig. 61-2 must be divided by the Ee of the generating plant that produced the electricity and the

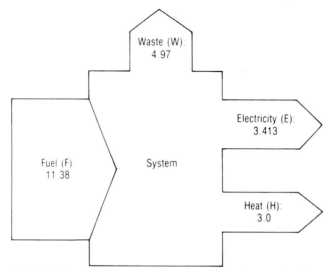

Fig. 61-3. First law balance diagram for electric/heat conversion "system."

quotient added to the thermal value of the input fuel for the thermal plant.

If the products of Figs. 61-1 and 61-2 were to be produced by an integrated plant producing both steam and electricity, the first law balance diagram would appear as in Fig. 61-3, where

$$e = 3.413 \text{ GB}$$

$$h = 3.0 \text{ GB}$$

$$f = 11.38 \text{ GB}$$

and

$$Ee = (e + h)/f$$
$$= (3.413 + 3.0)/11.38$$
$$= 0.56.$$

Note that the Ee of this combined plant, compared to the Ee of the two separate plants in serving the *same* community requirements (0.40),** reveals a 40 percent more effective use of the resources.

The Ee thus developed is analogous to the concept of seasonal efficiency as applied to either the first law or the second law plant. The energy-effectiveness factor makes it apparent that in an integrated plant, the "seasonal efficiency" for the generation of electric power

** The sum of the two outputs, 3.413 + 3.0, divided by the sum of the two inputs, 11.38 + 4.6, = 6.413/15.98 = 0.40.

is identical to the "seasonal efficiency" for the generation of thermal energy, and the "seasonal efficiency" for each is equal to the energy-effectiveness factor.

A look at heat removal

Most energy communities have three basic forms of product energy requirement:

• Electric energy—for lighting and power drives.
• Heat addition—to offset heat losses to surroundings maintaining comfort conditions, to heat water, and to provide for process needs.
• Heat removal—to remove heat and moisture gained from surroundings, to provide for low-temperature storage, and to meet low-temperature process needs.

The heat removal could be accomplished *within the community* by one or a combination of electric energy to power compression refrigeration machines, fuel to power a prime mover driving compression refrigeration or powering absorption refrigeration, or thermal energy to power absorption refrigeration. Even the most cursory consideration of the phenomenon of balancing thermal and shaft requirements of a combined plant reveals that the load balance is a function of the product delivered or, stated another way, a function of the *energy community systems*. If these are balanced between the thermal and electric forms such that they optimize the use of salvage heat from the second law generating process within the plant, the sum of $e + h$ compared to f will increase. As a result, the energy needs of the same community, although remaining fixed, will be provided for with less consumption of resource energy.

In many cases, such as central heating–cooling plants, total energy or cogeneration plants, and recent conceptual developments known as MIUS and TIES*** systems, the cooling requirements of the community are integrated into the plant. In this case, in addition to the electric and heat energy supplied to the community, heat energy is removed from the community to the plant through a low-temperature fluid system (either single or two phase). Fig. 61-4a is a first law balance diagram for a plant that includes products of electricity, heating, and cooling. Note in the first law balance that the cooling energy enters the plant boundaries. Despite this, it represents a plant *product* delivered. This phenomenon, recognized by any student of thermodynamics (that the energy flow is negative while the useful product is positive), is the phenomenon that led to such evaluation functions as coefficient of performance, COP, and more recently energy-efficiency ratio, EER, both conceived to express the relationship between low-temperature energy removed and energy consumed to accomplish the removal.

COP defined

Classically, coefficient of performance is defined as refrigeration effect divided by input energy required to accomplish the effect, both expressed in equivalent units.

As stated above, fuel, electricity, or thermal energy could be converted outside the boundaries of the plant to provide the cooling needs of the community. If this were done, the energy required would be:

***MIUS is Modular Integrated Utility System and TIES is Total Integrated Energy System.

$$e, h, \text{ or } f = \frac{\text{cooling required}}{(\text{COP})_s}$$

where

$(\text{COP})_s$ = COP of community refrigeration system(s).

It is germane to the concept of energy-effectiveness factor to recognize that *the consideration here is to establish the value of the product produced by the plant in units consistent with those of the other products, and that how the refrigeration is accomplished within the plant is irrelevant.* Thus, the numerical value of $(\text{COP})_s$ must be fixed at a level commercially available in community systems. Theoretically, the $(\text{COP})_s$ could be the annually integrated Carnot COP; in reality, however, the Carnot COP cannot be reasonably approached for any system being compared or evaluated.

As an example of a reasonable value relating to articles of commerce, it is suggested that $(\text{COP})_s$ be set at the values established in ASHRAE Standard 90, a consensus standard for energy conservation in new buildings. The standard sets minimum values of COP for various size categories of machinery, consistent with achievable limits in articles of commerce. Thus, these values, calculated by a weighted average of the connected loads, could serve as useful constants in defining a universal evaluation function.

The energy-effectiveness factor for a combined plant that includes cooling product can then be developed by addressing the difference between the first law balance diagram of Fig. 61-4a and the product diagram of Fig. 61-4b. Again, the boundaries of the system being evaluated must be carefully defined, and in the simplest terms must include the conversion plant, distribution system, and any satellite conversion apparatus connected to the distribution system. In converting from the first law balance concept to the product concept, *the boundaries must be held fixed.*

Referring to Fig. 61-4a, the energy values are:

E = annual electric energy output (kW-hr) $\times$ 3413,

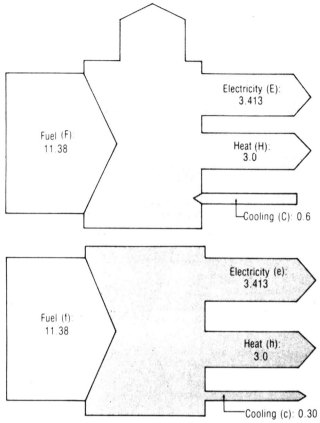

Fig. 61-4. a. First law balance diagram for electric/heating/cooling system. **b.** Product diagram for electric/heating/cooling system.

H = annual heat energy output (Btu),

C = annual cooling energy input (Btu),

F = annual fuel energy input (Btu),

W = annual energy wasted to environment (Btu).

And the product values for Fig. 61-4b are:

$$e = E,$$

$$h = H,$$

$$c = C/(COP)_s,$$

$$f = F.$$

The energy-effectiveness factor for the integrated plant of Fig. 61-4b is then:

$$Ee = (e + h + c)/f.$$

The numerical value of $(COP)_s$ used to convert the cooling energy of Fig. 61-4a to the cooling product of Fig. 61-4b was 1.8. This

value was taken from ASHRAE Standard 90, a 1977 value for apparatus with a capacity less than 65,000 Btuh, which would typically represent a residential-type energy community. The Ee for the combined plant serving the community is:

$$Ee = \frac{3.413 + 3.0 + 0.33}{11.38}$$

$$= 0.59.$$

Use *Ee* for solar, too

The concept of energy-effectiveness factor is a useful tool in quantifying the value of using nondepleting energy sources such as solar energy. As an example, if the "system" of Fig. 61-4 were to be designed with a solar collection and conversion system that, with the same product, required only 10 GB of fuel energy (reduced from 11.38 GB by the solar energy utilized), the Ee would be:

$$Ee = \frac{3.413 + 3.0 + 0.33}{10}$$

$$= 0.67.$$

Thus, there would be a 14 percent increase in the effectiveness of the use of resource energy resulting from the solar contribution.

In summary

Energy-effectiveness factor is based on the concept of *product value* ratio rather than energy ratio. This provides for the combination of cooling energy, heating energy, and electric energy in a single evaluation parameter (a combination required for a total evaluation of effectiveness in the use of energy in contemporary energy communities).

If properly applied, it provides a consistently valid evaluation function for both comparing alternative methods of conversion for any given energy community and comparing actual with anticipated performance using a minimum of instrumentation.

Energy-effectiveness factor can be applied to any integrated community, from a single residence to an entire living-working-transportation community served by a single conversion plant or by a multitude of plants and direct fuel supply points.

The fundamental requirements for proper application are to convert rigidly from energy to product terms and to define carefully the boundaries of the analysis.

In a study of one energy community served in 18 different manners, the energy-effectiveness factor was found to vary from 0.28 to 0.416 with the consumption of resource varying in inverse correlation from 360 to 263 GB per year.

Appendix
Acknowledgment of original publications

Chapter	Original Title	Originally Published	Date
1	Pure versus Applied Science	*Heating/Piping/Air Conditioning*	March 1977
2	The Design Process	*Heating/Piping/Air Conditioning*	March 1978
3	Design Parameters	*Heating/Piping/Air Conditioning*	Feb. 1975
4	Evaluation Functions	*Heating/Piping/Air Conditioning*	Dec. 1978
5	Fads: Their Influence on Design	*Heating/Piping/Air Conditioning*	Oct. 1978
6	Keep it Simple	*Heating/Piping/Air Conditioning*	May 1976
7	Specialty Devices (or Gadgets)	*Heating/Piping/Air Conditioning*	Feb. 1978
8	Minor Details	*Heating/Piping/Air Conditioning*	Sept. 1978
9	New Energy Technology: Closer Than We Think	*Heating/Piping/Air Conditioning*	May 1979
10	A Re-examination of Engineering Education	*Heating/Piping/Air Conditioning*	Jan. 1978
11	Education: The Key to the Energy Dilemma	*Heating/Piping/Air Conditioning*	Oct. 1976
12	The Professional's Role in Energy Management and Retrofit	*Specifying Engineer*	April 1978
13	Fifty Years of Contribution	*Heating/Piping/Air Conditioning*	Sept. 1979
14	Engineering: Can We Afford It?	*Heating/Piping/Air Conditioning*	April 1975
15	Facts About Energy	*Energy Management Handbook*	Dec. 1977
16	What Is Energy?	*Heating/Piping/Air Conditioning*	May 1975
17	Energy: A Unique Commodity	*Heating/Piping/Air Conditioning*	Dec. 1977
18	Energy Transportation	*Heating/Piping/Air Conditioning*	April 1976
19	Infinite Source	*Heating/Piping/Air Conditioning*	Feb. 1977
20	Section 12: Toward a More Effective Use of Energy Resources	*ASHRAE Journal*	May 1977
21	Energy Economics: A Needed Science	*Heating/Piping/Air Conditioning*	Jan. 1975
22	More On the Energy Hypothesis	*Heating/Piping/Air Conditioning*	Sept. 1975
23	Energy Economics: A Design Paramater	*Heating/Piping/Air Conditioning*	June 1973
24	Applied Science of Energy Economics in Building Systems	*ASHRAE Journal*	May 1974
25	Return to Regionalism in Building Design	*Heating/Piping/Air Conditioning*	Jan. 1976
26	Infinite Sink	*Heating/Piping/Air Conditioning*	Jan. 1979
27	Second Law Concepts	*Heating/Piping/Air Conditioning*	Feb. 1979
28	Energy Management	*Heating/Piping/Air Conditioning*	Jan. 1977
29	Building Automation Systems	*Heating/Piping/Air Conditioning*	Sept. 1977
30	The Laundry List	*Heating/Piping/Air Conditioning*	April 1979
31	Energy Audits I: The Energy Profile	*Heating/Piping/Air Conditioning*	July 1979
	Energy Audits II: Components of Energy Use	*Heating/Piping/Air Conditioning*	Aug. 1979
	Energy Audits III: Components of Energy Use	*Heating/Piping/Air Conditioning*	Oct. 1979
	Energy Audits IV: The Uses of Audits	*Heating/Piping/Air Conditioning*	Nov. 1979
32	A Primer on Electric Rates	*Heating/Piping/Air Conditioning*	March 1980
33	Local Building Codes	*Heating/Piping/Air Conditioning*	Dec. 1976
34	Standards I: The Cornerstone of Commerce	*Heating/Piping/Air Conditioning*	July 1978
	Standards II: The Role of Consensus Standards	*Heating/Piping/Air Conditioning*	Aug. 1978
35	The Computer and HVAC System Design	*Heating/Piping/Air Conditioning*	Oct. 1975

Appendix
Acknowledgment of original publications

Chapter	Original Title	Originally Published	Date
36	It's Time to Re-evaluate Computer Use	*Heating/Piping/Air Conditioning*	Jan. 1980
37	The Computer As a Tool for Energy Analysis	*Heating/Piping/Air Conditioning*	Jan. 1975
38	Computer Applications for Systems Design & Analysis*	*ASHRAE Transactions*	Vol. 82, Part II, 1976
39	Life Cycle Costing	*Heating/Piping/Air Conditioning*	Dec. 1975
40	Investment Optimization: A Methodology for Life Cycle Cost Analysis	*ASHRAE Journal*	Jan. 1977
41	A Single Equation for Co-generation Analyses	*Heating/Piping/Air Conditioning*	Dec. 1979
41	Extending the Co-generation Equation	*Heating/Piping/Air Conditioning*	Feb. 1980
42	Picking an Energy Source and Conversion System	*Heating/Piping/Air Conditioning*	Nov. 1975
43	The Myth of Free Steam	*Heating/Piping/Air Conditioning*	Nov. 1977
44	Oil Total Energy Plant Makes Turnpike Service Plaza Self-Sustaining	*ASHRAE Journal*	April 1969
45	The Status of Total Energy	*Heating/Piping/Air Conditioning*	May 1978
46	Hydronic Systems Overview: Part I	*Heating/Piping/Air Conditioning*	May 1977
	Hydronic Systems Overview: Part II	*Heating/Piping/Air Conditioning*	July 1977
	Hydronic Systems Overview: Part III	*Heating/Piping/Air Conditioning*	Aug. 1977
47	Integrated Decentralized Chilled Water Systems	*ASHRAE Transactions*	Vol. 82, Part I, 1976
48	Integrated Loop System for Campus Cooling	*Heating/Piping/Air Conditioning*	May 1973
49	Preheating Outdoor Air with Transfer Fluid Systems	*ASHRAE Transactions*	Jan. 1972
50	Steam: A State-of-the-Art Update	*Heating/Piping/Air Conditioning*	June 1975
50	Steam: A State-of-the-Art Update	*Heating/Piping/Air Conditioning*	Aug. 1975
51	The Key to Vapor Lock	*Heating/Piping/Air Conditioning*	March 1976
52	The Fan Curve: A Useful Tool?	*Heating/Piping/Air Conditioning*	April 1977
53	A New Look at Fan System Curves**	*Heating/Piping/Air Conditioning*	Nov. 1974
54	An Engineering Overview of System Balance and Energy Conservation	Purdue University (Ray W. Herrick Laboratories)	April 1976
55	Planned versus Breakdown Maintenance	*Heating/Piping/Air Conditioning*	March 1975
56	Look to the Obvious	*Heating/Piping/Air Conditioning*	July 1976
57	Realiability By Design	*Heating/Piping/Air Conditioning*	April 1978
58	Consumer and Commercial User Requirements As Seen By the Consulting Engineer	*ASHRAE Transactions*	June 1972
59	Thermodynamic versus System Efficiency	*Heating/Piping/Air Conditioning*	March 1979
60	Thermal Effectiveness of a Vapor Compression Cycle	*Heating/Piping/Air Conditioning*	Feb. 1976
61	Energy Effectiveness Factor	*Heating/Piping/Air Conditioning*	Aug. 1976

*Coauthored by Albert W. Black, III, and William J. Coad. **Coauthored by William J. Coad and Philip D. Sutherlin.

Index